W0105990

Lecture Notes
in Control and Information Sciences 189

Editors: M. Thoma and W. Wyner

Achim Ilchmann

Non-Identifier-Based
High-Gain Adaptive Control

Springer-Verlag London Ltd.

Author

Achim Ilchmann, PhD
Institut für Angewandte Mathematik, Universität Hamburg, Bundesstrasse 55,
20146 Hamburg, Germany

ISBN 978-3-540-19845-1   ISBN 978-3-540-39318-4 (eBook)
DOI 10.1007/978-3-540-39318-4

Preface

The seminal work of Morse, Nussbaum and Willems & Byrnes in 1983/4 has initiated the study of adaptive controllers for dynamical systems in which the adaptation strategy does not invoke any identification mechanism. Over the last decade, this field of adaptive control has become a major research topic. The present work gives a rather complete 'state of the art' of the following more specific area: The system classes under consideration contain linear (possibly nonlinearly perturbed), finite dimensional, continuous time systems which are stabilizable by high-gain output feedback, therefore, in particular the system is minimum phase. Simple adaptive controllers involving a simple switching strategy in the feedback are designed. The switching strategy is mainly tuned by a one parameter controller based on output data alone. Control objectives considered are stabilization, tracking, and servomechanism action. In addition, robustness with respect to nonlinear perturbations and performance improvements are investigated.

I wrote the present text during a two years research visit to the Centre for Systems and Control Engineering at the University of Exeter, U.K., from October 1990 - September 1992. The hospitality of the centre, with its stimulating environment, made a big contribution and I thank especially Dave Owens and Stuart Townley, of the centre, for numerous helpful discussions and suggestions. I also benefitted from many stimulating discussions with Gene Ryan of the University of Bath and Hartmut Logemann of the University of Bremen. I am indebted to Heinrich Voß of the University of Hamburg who read the manuscript and made several critical and helpful comments. Finally, thanks are due to Dieter Neuffer from the University of Stuttgart who spent considerable time and patience on introducing me to SIMULINK and MATLAB. My visit was made possible by a two years research grant from Deutsche Forschungsgemeinschaft (DFG) and additional support came from the University of Exeter and the EEC SCIENCE programme, which are hereby gratefully acknowledged.

Hamburg, February, 1993   *Achim Ilchmann*

Contents

Nomenclature

$\underline{N}$	$=$	$\{1, \ldots, N\}$
$\mathbb{R}_+ (\mathbb{R}_-)$		the set of non-negative (non-positive) real numbers
$\mathbb{C}_+ (\mathbb{C}_-)$		open right- (left-) half complex plane
$\mathbb{R}[s]^{m \times m}$		the set of $m \times m$ matrices over the ring of real polynomials
$\mathbb{R}(s)^{m \times m}$		the set of $m \times m$ matrices over the field of rational real functions
$GL_n(K)$		general linear group of invertible $n \times n$ matrices with entries in K
$B_\lambda(0)$	$=$	$\{x \in \mathbb{R}^m \mid \|x\| < \lambda\}$ for $m \in \mathbb{N}, \lambda > 0$
$\sigma(A)$		the spectrum of the matrix $A \in \mathbb{C}^{n \times n}$
$\mu_{\min}(A))$		minimal singular value of the matrix $A \in \mathbb{C}^{n \times m}$
$\mu_{\max}(A)$		maximal singular value of the matrix $A \in \mathbb{C}^{n \times m}$
$|A|$		the determinant of the matrix $A \in \mathbb{C}^{n \times n}$
$\|x\|_P$	$=$	$\sqrt{<x, Px>}$ for $x \in \mathbb{R}^n, P = P^T \in \mathbb{R}^{n \times n}$ positive-definite
$\|x\|$	$=$	$\|x\|_{I_n}$
$L_p(I)$		the vector space of measurable functions $f : I \to \mathbb{R}^n$, $I \subset \mathbb{R}$ an interval, n being defined by the context, such that $\|f(\cdot)\|_{L_p(I)} < \infty$, where

$$\|f(\cdot)\|_{L_p(I)} \;\;=\;\; \begin{cases} \left[\int_I \|f(s)\|^p \, ds\right]^{1/p} & \text{for} \quad p \in [1,\infty) \\[2mm] \operatorname*{ess\,sup}_{s \in I} \|f(s)\| & \text{for} \quad p = \infty \end{cases}$$

$\mathcal{C}^p(I, \mathbb{R}^m)$ — the vector space of p-times continuously differentiable functions $f : I \to \mathbb{R}^m$, $p \in \mathbb{N} \cup \{\infty\}$

$\mathcal{W}^{1,\infty}(I, \mathbb{R}^m)$ — the Sobolev space of functions $f : I \to \mathbb{R}^m$ which are absolutely continuous on compact intervals and $f(\cdot), \dot{f}(\cdot) \in L_\infty(\mathbb{R})$,

$d_\lambda(\cdot), D_\rho(\cdot)$ — 'distance' functions defined in (5.12) respectively (5.18).

Chapter 1

Introduction

A wide range of control theory deals with the design of a feedback controller for a *known* plant so that certain control objectives are achieved. The fundamental difference between this approach and that of *adaptive control* is that in adaptive control the plant is *not known* exactly, only structural information is available, e.g. minimality, minimum phase, or known relative degree. The aim is therefore to design a *single* controller which achieves prespecified control objectives for every member of a given class. The controller has to learn from the output data and, based on this information, to adjust its parameters.

The first attempts in adaptive control go back to the late 1950's, but it was only in the 1970's that a breakthrough was made. Subsequently, during the 1980's the field of adaptive control has matured. For a survey see Åström (1987) and Narendra (1991). Up to the end of the 1970's, adaptive controllers were a combination of identification or estimation mechanisms of the plant parameters together with a feedback controller. An area of *non-identifier-based* adaptive control was initiated by Mareels (1984), Mårtensson (1985), Morse (1983), Nussbaum (1983), and Willems and Byrnes (1984). In their approach, the adaptive feedback strategy is not based on any identification or estimation of the process to be controlled. This seminal work opened up an intensively studied specialised field within adaptive control, where the class of systems under consideration are either minimum phase or, more generally, only stabilizable and detectable. See Ilchmann (1991) for a survey.

In the present text, non-identifier-based adaptive controllers for minimum phase systems are studied, thus all controllers are designed according to the *high-gain* properties of the system class. No assumptions are made on the upper bound of the order of the process, nor on the upper bound of the sign of the high-frequency gain, no injection of probing signals is required, and the control strategy is more efficient than for non-minimum phase systems. The objective is to provide a single controller (consisting of a feedback law and a parameter adaptation law) which can control each system belonging to a certain class of systems. The control objectives are stabilization, tracking or servomechanism action, partly under performance requirements and in the presence of nonlinear perturbations.

We illustrate the idea by the simplest example we can think of: The system to be stabilized belongs to the class of scalar systems described by

$$\left.\begin{array}{rcl} \dot{x}(t) & = & ax(t) + bu(t) \quad , x(0) = x_0 \\ y(t) & = & cx(t) \end{array}\right\} \tag{1.1}$$

where $a, b, c, x_0 \in \mathbb{R}$ are unknown and the only structural assumption is $cb > 0$, i.e. the sign of the high-frequency gain is known to be positive.

If we apply the feedback law $u(t) = -ky(t)$ to (1.1), then the closed-loop system has the form

$$\dot{x}(t) = [a - kcb]x(t) \quad , x(0) = x_0. \tag{1.2}$$

Clearly, if $a/|cb| < |k|$ and $sign(k) = sign(cb)$, then (1.2) is exponentially stable. However, a, b, c are not known and thus the problem is to find adaptively an appropriate k so that the motion of the feedback system tends to zero.

Now a *time-varying* feedback is built into the feedback law

$$u(t) = -k(t)y(t), \tag{1.3}$$

where $k(t)$ has to be adjusted so that it gets large enough to ensure stability but also remains bounded. This can be achieved by the adaptation rule

$$\dot{k}(t) = y(t)^2, \qquad k(0) \in \mathbb{R}. \tag{1.4}$$

The nonlinear closed-loop system (1.1), (1.3), (1.4), i.e.

$$\dot{x}(t) = [a - k(t)cb]x(t), \quad k(t) = c^2 \int_0^t x(s)^2 \, ds + k(0), \quad (k(0), x(0)) \in \mathbb{R}^2 \tag{1.5}$$

has at least a solution on a small interval $[0, \omega)$, and the non-trivial solution

$$x(t) = e^{\int_0^t [a - k(s)cb] ds} x(0)$$

is monotonically increasing as long as $a - k(t)cb > 0$. Hence $k(t) \geq t(cx(0))^2 + k(0)$ increases as well. Therefore, there exists a $t^* \geq 0$ such that $a - k(t^*)cb = 0$ and (1.5) yields $a - k(t)cb < 0$ for all $t > t^*$. Hence the solution $x(t)$ decays exponentially and $\lim_{t \to \infty} k(t) = k_\infty \in \mathbb{R}$ exists. This is a special example for the following concept of universal adaptive control.

Suppose Σ denotes a certain class of linear, finite dimensional, time-invariant systems of the form

$$\left.\begin{array}{rcl} \dot{x}(t) & = & Ax(t) + Bu(t) \quad , x(0) = x_0 \\ y(t) & = & Cx(t) + Du(t) \end{array}\right\} \tag{1.6}$$

where $(A, B, C, D) \in \mathbb{R}^{n \times n} \times \mathbb{R}^{n \times m} \times \mathbb{R}^{m \times n} \times \mathbb{R}^{m \times m}$ are unknown, m is usually fixed, the state dimension n is an arbitrary and unknown number. The aim is to design a single adaptive output feedback mechanism $u(t) = \mathcal{F}\left(y(\cdot)|_{[0,t]}\right)$ which is a universal stabilizer for the class Σ, i.e. if $u(t) = \mathcal{F}\left(y(\cdot)|_{[0,t]}\right)$ is applied to any system (1.6) belonging to Σ, then the output $y(t)$ of the closed-loop system tends to zero as t tends to infinity and the internal variables are bounded.

In the present text, most of the adaptive stabilizers are of the following simple form (cf. Figure 1.1): A 'tuning' parameter $k(t)$, generated by an adaptation law

$$\dot{k}(t) = g(y(t)), \qquad k(0) = k_0, \tag{1.7}$$

where $g : \mathbb{R}^m \rightarrow \mathbb{R}$ is continuous and locally Lipschitz, is implemented into the feedback law via

$$u(t) = F(k(t))y(t), \tag{1.8}$$

where $F : \mathbb{R} \rightarrow \mathbb{R}^{m \times m}$ is piecewise continuous and locally Lipschitz.

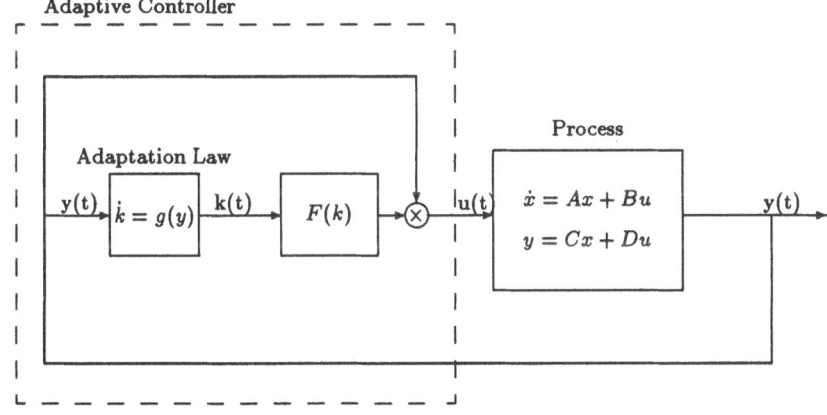

Figure 1.1: Universal adaptive stabilizer

Definition 1.1.1

A controller, consisting of the adaptation law (1.7) and the feedback rule (1.8), is called a *universal adaptive stabilizer* for the class of systems Σ, if for arbitrary initial condition $x_0 \in \mathbb{R}^n$ and any system (1.6) belonging to Σ, the closed-loop system (1.6)-(1.8) has a solution the properties

(i) there exists a unique solution $(x(\cdot), k(\cdot)) : [0, \infty) \rightarrow \mathbb{R}^{n+1}$,

(ii) $x(\cdot), y(\cdot), u(\cdot), k(\cdot)$ are bounded,

(iii) $\lim_{t \to \infty} y(t) = 0$,

(iv) $\lim_{t \to \infty} k(t) = k_\infty \in \mathbb{R}$ exists.

The concept of adaptive *tracking* is similar. Suppose a class $\mathcal{Y}_{\text{ref}}$ of reference signals is given. It is desired that the error between the output $y(t)$ of (1.6) and the reference signal $y_{\text{ref}}(t)$

$$e(t) := y(t) - y_{\text{ref}}(t)$$

is forced, via a simple adaptive feedback mechanism, either to zero or towards a ball around zero of arbitrary small prespecified radius $\lambda > 0$. The latter is called λ-tracking. To achieve asymptotic tracking, an internal model

$$\left.\begin{array}{rcl} \dot{\xi}(t) & = & A^* \xi(t) + B^* v(t) \quad , \xi(0) = \xi_0 \\ u(t) & = & C^* \xi(t) + D^* v(t) \end{array}\right\} \tag{1.9}$$

where $(A^*, B^*, C^*, D^*) \in \mathbb{R}^{n' \times n'} \times \mathbb{R}^{n' \times m} \times \mathbb{R}^{m \times n'} \times \mathbb{R}^{m \times m}$, is implemented in series interconnection with an universal adaptive stabilizer, cf. Figure 1.2. The precompensator resp. internal model (1.9) contains the dynamics of the reference signals. An internal model is not necessary if λ-tracking is desired.

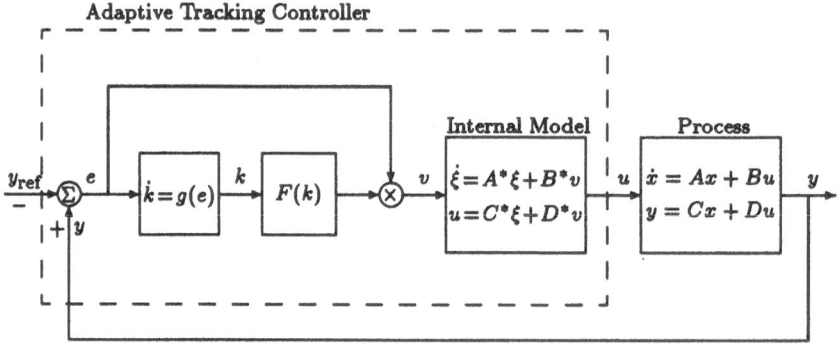

Figure 1.2: Universal Adaptive Tracking with Internal Model

Definition 1.1.2

A controller, consisting of an *adaptation law* (1.7), a *feedback law* (1.8), and an internal model (1.9) is called a *universal adaptive tracking controller* for the class of systems Σ and reference signals $\mathcal{Y}_{\text{ref}}$, if for every $y_{\text{ref}}(\cdot) \in \mathcal{Y}_{\text{ref}}$, $x_0 \in \mathbb{R}^n$, $\xi_0 \in \mathbb{R}^{n'}$, and every system (1.6) belonging to Σ, the closed-loop system (1.6)-(1.9) satisfies

(i) there exists a unique solution $(x(\cdot), \xi(\cdot), k(\cdot)) : [0, \infty) \to \mathbb{R}^{n+n'+1}$,

(ii) the variables $x(t), y(t), u(t), \xi(t)$ 'blow up' no faster than $y_{\text{ref}}(t)$,

(iii) $\lim_{t \to \infty} [y(t) - y_{\text{ref}}(t)] = 0$,

(iv) $\lim_{t \to \infty} k(t) = k_\infty \in \mathbb{R}$ exists.

For prespecified $\lambda > 0$, (1.7), (1.8) is called a *universal adaptive λ-stabilizer* resp. *λ-tracking controller* if (i),(ii), and (iv) hold true, but instead of (iii) the weaker condition

$$[y(t) - y_{\text{ref}}(t)] \to \overline{B}_\lambda(0) \quad \text{as} \quad t \to \infty$$

is satisfied, i.e. the error $e(t)$ approaches the closed ball of radius λ around zero as t tends to ∞.

If a universal adaptive controller is applied to (1.6) for some $(x_0, k_0) \in \mathbb{R}^n \times \mathbb{R}$, then we call the system

$$\dot{x}(t) = \left[A + BF(k_\infty)(I_m + DF(k_\infty))^{-1} C \right] x(t)$$

the *terminal* system, provided it is well defined.

Many results presented in this text fit into the framework described above. We will also consider linear systems subjected to nonlinear perturbations in the state, input and output, corrupted input and output noise, and non-differentiable gain adaptation. Due to the nonlinearities, the solution of the closed-loop system of the nominal system and the adaptive feedback mechanism is no longer unique, but all solution will meet the desired control objectives. Some feedback strategies contain non-differential gain adaptation, other have *multiparameter* gain adaptation, i.e. $g(\cdot)$ in (1.7) is a mapping from $\mathbb{R}^m$ to $\mathbb{R}^{m'}$ for some $m' > 1$. Many results can be extended, by no means trivially, to classes of infinite-dimensional systems. This will not be treated here, but the available literature is discussed in the sections 'Notes and References' ending each chapter.

The outline of the chapters is as follows.
In Chapter 2 and 3 we do not deal with any adaptation mechanisms, instead the system classes under consideration are analysed and results to be used later are prepared.
In Chapter 2, we study in detail the properties of multivariable minimum phase systems, give convenient state space forms, prove the so called high-gain lemmata for relative degree 1 and 2 systems, and derive important inequalities relating past inputs and outputs with the present output. This leads to a rather complete understanding of high-gain stabilizable systems and tools which

will be used throughout the remaining chapters.

In Chapter 3, the class of strictly positive real and almost strictly positive real systems is investigated. Although this class is more restrictive than minimum phase systems it turns out that, by simple input and output transformations, every minimum phase system is equivalent to an almost strictly positive real system. The result will become important for the stability proofs of adaptive stabilizers given in Chapter 4. Recently published results on multivariable strictly positive real systems, in particular its relationsphip to the Lur'e equations, are used to understand the effect of high-gain control, and simplify the proofs for adaptive stabilizers. Moreover, it allows consideration not only of strictly proper, but of proper linear systems.

In the first section of Chapter 4, the concept of switching functions is studied. Different switching functions are introduced and it is shown how they relate to the Nussbaum conditions. These results are needed for the remaining sections of the chapter. In Section 4.2, various universal adaptive stabilizers based on Nussbaum-type switching are derived for systems (A, B, C, D) under mild assumptions on the high-frequency gain matrix. A universal adaptive stabilizer for the class of all single-input, single-output, positive or negative high-gain stabilizable systems is given. Throughout the section the feedback strategy is continuous respectively piecewise continuous, i.e. $F(\cdot)$ in (1.8) is continuous respectively piecewise constant. In Section 4.3, an alternative switching strategy based on a switching decision function is introduced. Apart from robustness properties shown in Chapter 6, the advantage of this different approach is that there is no need to implement a scaling-invariant Nussbaum function which is behaving very rapidly. In Sections 4.4 and 4.5, we deal with the problem of how to obtain exponential decay of the state $x(t)$ of the closed-loop system. In Section 4.4, this is achieved by introducing an exponential weighting factor in the gain adaptation, whereas in Section 4.5 a different approach uses piecewise constant gain implementation in the feedback strategy.

In Chapter 5, the problem of how to track signals belonging to a class of reference signals is investigated. One solution to the problem of asymptotic tracking is presented in Section 5.1. We use series interconnection between an internal model, representing the dynamics of the reference signals, and universal adaptive stabilizers studied in Chapter 4. This approach is restricted to sinusoid reference signals, whereas in Section 5.2, at the expense of λ-tracking, a controller using dead-zones is introduced which does not invoke an internal model and works for a much larger class of reference signals.

Well posedness and robustness properties of the universal adaptive stabilizers of Chapter 4 and the asymptotic and λ-tracking controllers of Chapter 5 are investigated in Chapter 6. In Section 6.1, it is proved that the problem of universal adaptive stabilization is well posed with respect to nonlinearities in the state equation of the nominal system. Robustness with respect to other nonlinear perturbations is proved. In section 6.2, we show that many universal adaptive stabilizers of single-input, single-output systems tolerate sector bounded input-output nonlinearities, for multi-input, multi-output almost strictly positive real systems ever multivariable sector bounded input-output nonlinearities are allowed. In Section 6.3, we prove that the λ-stabilization respectively λ-tracking controller is capable of tolerating a much larger class of nonlinear perturbations in the input and state as well as sector bounded input-output nonlinearities, even input and output corrupted noise is allowed.

The purpose of Section 7.1. is to illustrate the qualitative dynamical behaviour of many universal adaptive stabilizers and to introduce modifications of the previous universal adaptive stabilizers and λ-tracking controllers which lead to an improvement of the transient behaviour. In Section 7.2, a universal adaptive stabilizer is designed which, at the expense of derivative feedback, achieves (prespecified) arbitrarily small overshoot of the output, and, moreover, guarantees that the output is less than an arbitrarily small, prespecified constant in an arbitrarily small, prespecified period of time.

The results on root-loci of single-input, single-output minimum phase systems derived in Section 8.1 are used in Section 8.2 to show that that the piecewise constant stabilizer introduced in Section 4.5 'almost always' (w.r.t. the sequence of thresholds) yields an exponentially stable terminal system.

Each chapter is finalized with a section on the literature and related problems, in particular, extended results for infinite-dimensional systems are quoted.

Chapter 2

High-Gain Stabilizability

In this chapter, we derive several properties of high-gain stabilizable and/or minimum phase systems which are essential for getting a deeper insight into these system classes, and which will be used throughout the remaining chapters.

2.1 Minimum phase systems

We shall show that all non-trivial systems which are stabilizable by high-gain output feedback are necessarily minimum phase. Therefore, it is important to study the class of minimum phase systems. We will also give simple and convenient state space forms for relative degree 1 or 2 systems and prove a crucial integral inequality relating past input and output data to the present output.

Definition 2.1.1
Let $G(\cdot) \in \mathbb{R}(s)^{m \times m}$ be a rational matrix with *Smith-McMillan form*

$$\text{diag}\left\{ \frac{\varepsilon_1(s)}{\psi_1(s)}, \ldots, \frac{\varepsilon_r(s)}{\psi_r(s)}, 0, \ldots, 0 \right\} = U(s)^{-1} G(s) V(s)^{-1} \qquad (2.1)$$

where $U(\cdot), V(\cdot) \in \mathbb{R}[s]^{m \times m}$ are unimodular, $rk_{\mathbb{R}(s)} G(\cdot) = r$, $\varepsilon_i(\cdot), \psi_i(\cdot) \in \mathbb{R}[s]$ are monic and coprime and satisfy $\varepsilon_i(\cdot)|\varepsilon_{i+1}(\cdot)$, $\psi_{i+1}(\cdot)|\psi_i(\cdot)$ for $i = 1, \ldots, r$. Set

$$\varepsilon(s) := \prod_{i=1}^{r} \varepsilon_i(s), \qquad \psi(s) := \prod_{i=1}^{r} \psi_i(s).$$

s_0 is a *(transmission) zero* of $G(\cdot)$, if $\varepsilon(s_0) = 0$, and a *pole* of $G(\cdot)$, if $\psi(s_0) = 0$. If $G(\cdot) = g(\cdot) \in \mathbb{R}[s]$, then $\deg \psi(\cdot) - \deg \varepsilon(\cdot)$ is called the *relative degree* of $g(\cdot)$.

$G(\cdot)$ is *proper* resp. *strictly proper* if deg $\psi(\cdot) \geq$ deg $\varepsilon(\cdot)$ resp. deg $\psi(\cdot) >$ deg $\varepsilon(\cdot)$.
The system

$$
\begin{aligned}
\dot{x}(t) &= Ax(t) + Bu(t), \\
y(t) &= Cx(t) + Du(t)
\end{aligned}
$$

with $(A, B, C, D) \in \mathbb{R}^{n \times n} \times \mathbb{R}^{n \times m} \times \mathbb{R}^{m \times n} \times \mathbb{R}^{m \times m}$, is called a *minimal realization* of $G(\cdot) \in \mathbb{R}(s)^{m \times m}$, if (A, B) is controllable and (A, C) is observable and $G(s) = C(sI_n - A)^{-1}B + D$.
$G(\cdot)$ is said to be *minimum phase*, if

$$
\varepsilon(s) \neq 0 \quad \text{for all} \quad s \in \overline{\mathbb{C}}_+.
$$

A state space system $(A, B, C, D) \in \mathbb{R}^{n \times n} \times \mathbb{R}^{n \times m} \times \mathbb{R}^{m \times n} \times \mathbb{R}^{m \times m}$ is called *minimum phase*, if it is stabilizable and detectable and $G(s)$ has no zeros in $\overline{\mathbb{C}}_+$.

A characterization of the minimum phase condition for state space system is given in the following proposition.

Proposition 2.1.2
$(A, B, C, D) \in \mathbb{R}^{n \times n} \times \mathbb{R}^{n \times m} \times \mathbb{R}^{m \times n} \times \mathbb{R}^{m \times m}$ satisfies

$$
det \begin{bmatrix} sI_n - A & -B \\ -C & -D \end{bmatrix} \neq 0 \quad \text{for all} \quad s \in \overline{\mathbb{C}}_+
$$

if, and only if, the following three conditions are satisfied

(i) $rk\,[sI_n - A, B] = n$ for all $s \in \overline{\mathbb{C}}_+$,   i.e. (A, B) is stabilizable by state feedback,

(ii) $rk \begin{bmatrix} sI_n - A \\ C \end{bmatrix} = n$ for all $s \in \overline{\mathbb{C}}_+$,   i.e. (A, C) is detectable,

(iii) $G(s)$ has no zeros in $\overline{\mathbb{C}}_+$.

Proof: We use the notation of Definition 2.1.1. Coppel (1974), Theorem 10, has proved that, if (A, B, C) is detectable and stabilizable, then $s_0 \in \overline{\mathbb{C}}_+$ is a zero of $\psi(\cdot)$ (including multiplicity) if and only if it is a zero of $det(\cdot I_n - A)$. Using this result together with Schur's formula, see e.g. Gantmacher (1959), the proposition follows from

$$
\begin{vmatrix} sI_n - A & -B \\ -C & -D \end{vmatrix} = |sI_n - A| \cdot |-D - C(sI_n - A)^{-1}B| = \frac{|sI_n - A|}{\psi(s)} \cdot \varepsilon(s).
$$

$\square$

The following lemma provides a useful state space form into which every system

$$
\left.
\begin{aligned}
\dot{x}(t) &= A x(t) + B u(t), \qquad x(0) = x_0 \in \mathbb{R}^n \\
y(t) &= C x(t),
\end{aligned}
\right\} \tag{2.2}
$$

with $(A, B, C) \in \mathbb{R}^{n \times n} \times \mathbb{R}^{n \times m} \times \mathbb{R}^{m \times n}$ with $det(CB) \neq 0$ can be converted. The state space transformation is representing the direct sum of the range of B and the kernel of C. It makes possible the separation of the inputs and outputs from the rest of the system states.

Lemma 2.1.3
Consider the system (2.2) with $det(CB) \neq 0$ and let $V \in \mathbb{R}^{n \times (n-m)}$ denote a basis matrix of $ker\, C$. It follows that $S := [B(CB)^{-1}, V]$ has the inverse $S^{-1} = [C^T, N^T]^T$, where $N := (V^T V)^{-1} V^T [I_n - B(CB)^{-1} C]$. Hence the state space transformation $(y^T, z^T)^T = S^{-1} x = ((Cx)^T, (Nx)^T)^T$ converts (2.2) into

$$
\left.
\begin{aligned}
\dot{y}(t) &= A_1 y(t) + A_2 z(t) + CB u(t) \\
\dot{z}(t) &= A_3 y(t) + A_4 z(t)
\end{aligned}
\right., \qquad
\begin{pmatrix} y(0) \\ z(0) \end{pmatrix} = S^{-1} x_0. \right\} \tag{2.3}
$$

Here $A_1 \in \mathbb{R}^{m \times m}$, $A_2 \in \mathbb{R}^{m \times (n-m)}$, $A_3 \in \mathbb{R}^{(n-m) \times m}$, $A_4 \in \mathbb{R}^{(n-m) \times (n-m)}$, so that

$$
\begin{bmatrix} A_1 & A_2 \\ A_3 & A_4 \end{bmatrix} = S^{-1} A S.
$$

If (A, B, C) is minimum phase, then A_4 in (2.3) is asymptotically stable.

Proof: The proof of the transformation is straightforward and therefore omitted. Stability of A_4 is a consequence of $det(CB) \neq 0$, the minimum phase assumption, and of the following equation which holds for all $s \in \overline{\mathbb{C}}_+$

$$
\begin{vmatrix} sI_n - A & B \\ C & 0 \end{vmatrix} = \begin{vmatrix} sI_m - A_1 & -A_2 & CB \\ -A_3 & sI_{n-m} - A_4 & 0 \\ I_m & 0 & 0 \end{vmatrix} = |CB| \cdot |sI_{n-m} - A_4| \neq 0.
$$

$\square$

For strictly proper, scalar, minimum phase systems of relative degree 2, a slightly more complicated but still very useful state space description is also available.

Lemma 2.1.4
If the system (2.2) is single-input, single-output, $(A, B, C) = (A, b, c)$, and of relative degree 2, i.e. $cb = 0$, $cAb \neq 0$, then there exists a coordinate transformation $S \in GL_n(\mathbb{R})$, such that $S^{-1}x = (y, \dot{y}, z^T)^T$ converts (2.2) into

$$
\frac{d}{dt}\begin{pmatrix} y(t) \\ \dot{y}(t) \\ z(t) \end{pmatrix} = \begin{bmatrix} 0 & 1 & 0 \\ a_2 & \frac{cA^2b}{cAb} & a_4^T \\ a_5 & 0 & A_6 \end{bmatrix}\begin{pmatrix} y(t) \\ \dot{y}(t) \\ z(t) \end{pmatrix} + \begin{bmatrix} 0 \\ cAb \\ 0 \end{bmatrix} u(t) \quad (2.4)
$$

where $a_2 \in \mathbb{R}$, $a_4, a_5 \in \mathbb{R}^{n-2}$, $A_6 \in \mathbb{R}^{(n-2)\times(n-2)}$.
If (2.2) is minimum phase, then $\sigma(A_6) \subset \mathbb{C}_-$.

Proof: Choose $V \in \mathbb{R}^{n\times(n-2)}$ of full rank so that $\ker \begin{bmatrix} c \\ cA \end{bmatrix} = V \cdot \mathbb{R}^{n-2}$.
It is easily verified, that the inverse of $S_1 := [Ab, b, V](cAb)^{-1}$ is given by

$$
S_1^{-1} = \begin{bmatrix} c \\ cA - \frac{cA^2b}{cAb}c \\ cAbV^*\left[I_n - \left(Ab - \frac{cA^2b}{cAb}b\right)(cAb)^{-1}c - bcA(cAb)^{-1}\right] \end{bmatrix},
$$

$V^* := (V^TV)^{-1}V^T$, and hence

$$
\bar{c} := cS_1 = [1, 0, \ldots, 0], \quad \bar{b} := S_1^{-1}b = \begin{bmatrix} 0 \\ cAb \\ 0 \end{bmatrix}, \quad \bar{A} := S_1^{-1}AS_1 = \begin{bmatrix} * & 1 & * \\ * & * & * \\ * & * & * \end{bmatrix},
$$

and $\bar{c}\bar{A}\bar{b} = cAb$. Applying elementrary row and columm operations of the form

$$
S_2^{-1} := \begin{bmatrix} 1 & 0 & 0 \\ 0 & 1 & 0 \\ * & 0 & I_{n-2} \end{bmatrix} \cdot \begin{bmatrix} 1 & 0 & 0 \\ * & 1 & * \\ 0 & 0 & I_{n-2} \end{bmatrix}
$$

to $\bar{A}$ yields

$$
\hat{A} := (S_1S_2)^{-1}AS_1S_2 = \begin{bmatrix} 0 & 1 & 0 \\ a_2 & a_3 & a_4^T \\ a_5 & 0 & A_6 \end{bmatrix}.
$$

Moreover, $\bar{c}S_2 = \bar{c}$, $S_2^{-1}\bar{b} = \bar{b}$, and $cA^2b = \bar{c}\hat{A}^2\bar{b} = a_3cAb$, and hence (2.4) holds with $S = S_1S_2$. Using (2.4), we have

$$
\begin{vmatrix} \lambda I_n - A & b \\ c & 0 \end{vmatrix} = \begin{vmatrix} \lambda & -1 & 0 & 0 \\ -a_2 & \lambda - a_3 & a_4^T & cAb \\ -a_5 & 0 & \lambda I_{n-2} - A_6 & 0 \\ 1 & 0 & 0 & 0 \end{vmatrix} = -cAb\cdot|\lambda I_{n-2} - A_6| .
$$

If (2.2) is minimum phase, then it follows that $|\lambda I_{n-2} - A_6| \neq 0$ for all $\lambda \in \overline{\mathbb{C}}_+$, and hence A_6 is asymptotically stable. This completes the proof. □

Another important consequence of strictly proper minimum phase systems with $\lim_{s \to \infty} sG(s) \in GL_m(\mathbb{R})$ (for single-input single-output this simply means they are of relative degree 1) is, that a simple input-output description of the system is possible.

Lemma 2.1.5
If the system (2.2) is minimum phase with $det(CB) \neq 0$, then there exists a bounded and causal operator

$$\mathcal{L} : L_p(0, \infty) \rightarrow L_p(0, \infty), \quad \text{for all} \quad p \in [1, \infty],$$

so that the input-output behaviour of (2.2) is described by

$$\dot{y}(t) = A_1 y(t) + \mathcal{L}(y(\cdot))(t) + CBu(t) + w(t), \quad y(0) = Cx_0 \qquad (2.5)$$

with $A_1 \in \mathbb{R}^{m \times m}$, and $w(\cdot) : [0, \infty) \rightarrow \mathbb{R}^m$ an exponentially decaying analytic function taking into account the initial condition of a part of the internal state.

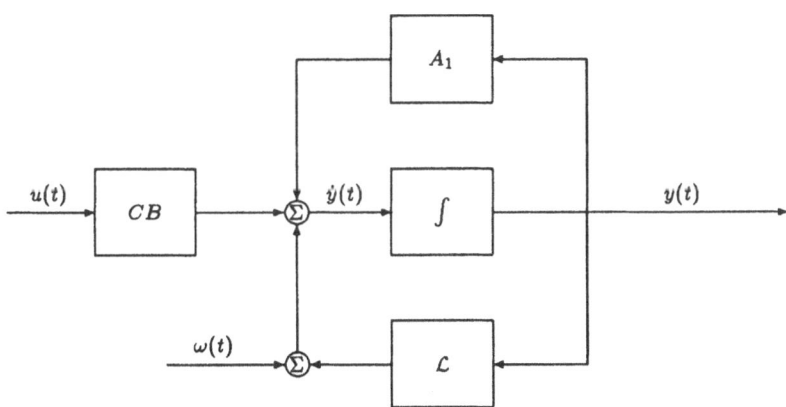

Figure 2.1: Input-Output Description

Proof: Without loss of generality we may assume that the system (2.2) is in the form (2.3). Define the causal operator

$$\mathcal{L}(y(\cdot))(t) := A_2 \int_0^t e^{A_4(t-s)} A_3 y(s) ds.$$

Since A_4 is asymptotically stable, there exist $M_1, \varepsilon > 0$, so that

$$\|e^{A_4 t}\| \leq M_1 e^{-\varepsilon t} \quad \text{for all} \quad t \geq 0.$$

This yields, see e.g. Vidyasagar (1978) pp. 250-254, that

$$\|\mathcal{L}(y)\,(\cdot)\|_{L_p(0,t)} \leq \frac{M_1\|A_2\|\|A_3\|}{\varepsilon} \|y(\cdot)\|_{L_p(0,t)} \quad \text{for all} \quad t \geq 0. \qquad (2.6)$$

Therefore, $\mathcal{L} : L_p(0, \infty) \rightarrow L_p(0, \infty)$ is well defined and bounded. By applying Variations of Constants to the second equation in (2.3), and inserting $z(t)$ into the first equation, we obtain

$$\dot{y}(t) = A_1 y(t) + A_2 \left[e^{A_4 t} z(0) + \int_0^t e^{A_4(t-s)} A_3 y(s) ds \right] + CBu(t).$$

Setting $w(t) = A_2 e^{A_4 t} z(0)$, the proof of the lemma is complete. $\square$

For strictly proper minimum phase systems (A, B, C) with $det(CB) \neq 0$, it is possible to relate the present output to past output and input data via the following inequality, where no information of the state variables is required. This will be an another important tool for the stability proofs of universal adaptive stabilizers in Chapter 4.

Lemma 2.1.6
Suppose the system (2.2) is minimum phase and satisfies $det(CB) \neq 0$. Let

$$\beta(\cdot) : \mathrm{I\!R}^m \rightarrow \mathrm{I\!R}^m, \qquad y \mapsto \beta(y) = \begin{cases} \frac{y}{\|y\|_P} & , \quad \text{if} \quad y \neq 0 \\ 0 & , \quad \text{if} \quad y = 0. \end{cases}$$

Then for every positive-definite matrix $P = P^T \in \mathrm{I\!R}^{m \times m}$ there exists $M > 0$ (depending only on A, B, C and P) such that

$$\frac{1}{2}\|y(t)\|_P^2 \leq M\|x_0\|^2 + M \int_0^t \|y(s)\|^2 ds + \int_0^t \langle y(s), PCBu(s)\rangle ds,$$

or, more general, for arbitrary $p \geq 1$, it holds that

$$\frac{1}{p}\|y(t)\|_P^p \leq M\|x_0\|^p + M \int_0^t \|y(s)\|^p ds + \int_0^t \|y(s)\|_P^{p-1} \langle \beta(y(s)), PCBu(s)\rangle ds$$

$$(2.7)$$

for arbitrary initial condition $x(0) = x_0 \in \mathrm{I\!R}^n$, for arbitrary piecewise continuous $u(\cdot) : [0, \omega) \rightarrow \mathrm{I\!R}^m$, where $\omega \in (0, \infty]$, and for all $t \in [0, \omega)$.

Proof: (a): We first consider the case $p = 2$. (2.5) yields, for all $s \in [0, \omega)$,

$$\frac{1}{2}\frac{d}{ds}(\|y(s)\|_P^2) = \langle y(s), PA_1 y(s) + Pw(s) + P\mathcal{L}(y)(s) + PCBu(s)\rangle$$

$$\leq M_2\|y(s)\|^2 + M_2\|y(s)\| (\|w(s)\| + \|\mathcal{L}(y)(s)\|)$$

$$+\langle y(s), PCBu(s)\rangle \tag{2.8}$$

for $M_2 := \|PA_1\| + \|P\|$. Applying Hölder's inequality and using (2.6) gives

$$\int_0^t \|y(s)\| \cdot \|\mathcal{L}(y)(s)\| ds \leq \left(\int_0^t \|y(s)\|^2 ds\right)^{\frac{1}{2}} \left(\int_0^t \|\mathcal{L}(y)(s)\|^2 ds\right)^{\frac{1}{2}}$$

$$= \sqrt{\frac{M_1 \|A_2\| \|A_3\|}{\varepsilon}} \int_0^t \|y(s)\|^2 ds \tag{2.9}$$

and, since $w(s) = A_2 e^{A_4 s} z(0)$,

$$\int_0^t \|y(s)\| \cdot \|w(s)\| ds \leq \left(\int_0^t \|y(s)\|^2 ds\right)^{\frac{1}{2}} M_3 \|z(0)\|$$

$$\leq \int_0^t \|y(s)\|^2 ds + M_3^2 \|z(0)\|^2 \tag{2.10}$$

for

$$M_3 := \sup_{t \geq 0} \left(\int_0^t \|A_2 e^{A_4 s}\|^2 ds\right)^{\frac{1}{2}}.$$

Integration of (2.8) over $[0, t]$ and inserting (2.9) and (2.10) yields, for all $t \in [0, \omega)$,

$$\frac{1}{2}\|y(t)\|_P^2 \leq \frac{1}{2}\|y(0)\|_P^2 + M_2 M_3^2 \|z(0)\|^2 + \int_0^t \langle y(s), PCBu(s)\rangle ds$$

$$+M_2 \left(2 + \sqrt{\frac{M_1 \|A_2\| \|A_3\|}{\varepsilon}}\right) \int_0^t \|y(s)\|^2 ds.$$

This proves the first inequality.
(b): The map $t \mapsto \|y(t)\|_P$ is not differentiable but, since $y(\cdot)$ is differentiable, it is absolutely continuous. Therefore, the set

$$J_1 := \{t \in [0, \omega) \mid \|y(t)\|_P \text{ is not differentiable}\}$$

is of measure zero. Let

$$J_2 := \{t \in [0, \omega) \mid u(\cdot) \text{ is not continuous at } t\}.$$

Now a routine calculation gives, for all $s \in \mathbb{R}_+ \setminus (J_1 \cup J_2)$,

$$\frac{d}{ds}\|y(s)\|_P = \begin{cases} \dfrac{\langle y(s), P\dot{y}(s)\rangle}{\|y(s)\|_P} & , y(s) \neq 0 \\ 0 & , y(s) = 0. \end{cases}$$

Thus it follows from (2.5) that, for all $s \in \mathbb{R}_+ \setminus (J_1 \cup J_2)$,

$$\frac{1}{p}\frac{d}{ds}\left(\|y(s)\|_P^p\right) = \|y(s)\|_P^{p-1}\langle\beta(y(s)), PA_1 y(s)$$

$$+Pw(s) + P\mathcal{L}(y)(s) + PCBu(s)\rangle$$

$$\leq M_4\|\beta(y(s))\|\left[\|y(s)\|^p + \|\mathcal{L}(y)(s)\|\|y(s)\|^{p-1}\right.$$

$$\left.+\|y(s)\|^{p-1}\|w(s)\|\right]$$

$$+\|y(s)\|_P^{p-1}\langle\beta(y(s)), PCBu(s)\rangle,$$

where

$$M_4 := \|P\|^{\frac{p-1}{2}}\left[\|PA_1\| + \|P\|\right].$$

An application of Hölder's inequality gives, for $q = \frac{p}{p-1}, \frac{1}{q} + \frac{1}{p} = 1$ and every $v(\cdot) \in L_p(0, t)$,

$$\int_0^t \|y(s)\|^{p-1}\|v(s)\|ds \leq \left[\int_0^t \|y(s)\|^{(p-1)q}ds\right]^{1/q} \cdot \left[\int_0^t \|v(s)\|^p ds\right]^{1/p}$$

$$= \|y(\cdot)\|_{L_p(0,t)}^{p-1}\|v(\cdot)\|_{L_p(0,t)}. \qquad (2.11)$$

Since $J_1 \cup J_2$ is of measure zero, integration of $\frac{d}{ds}\|y(s)\|_P^p$ over $[0, t] \setminus (J_1 \cup J_2)$ yields, by using (2.11), for all $t \in [0, \omega)$,

$$\frac{1}{p}\|y(t)\|_P^p \leq \frac{1}{p}\|y(0)\|_P^p + \int_0^t \|y(s)\|^{p-1}\langle\beta(y(s)), PCBu(s)\rangle ds$$

$$+M_5\int_0^t \|y(s)\|^p + \|y(s)\|^{p-1}\left[\|\mathcal{L}(y)(s)\| + \|w(s)\|\right]ds$$

$$\leq \frac{1}{p}\|y(0)\|_P^p + \int_0^t \|y(s)\|^{p-1}\langle\beta(y(s)), PCBu(s)\rangle ds$$

$$+M_5\left[\|y(\cdot)\|_{L_p(0,t)}^p + \|y(\cdot)\|_{L_p(0,t)}^{p-1}\|\mathcal{L}(y)(\cdot)\|_{L_p(0,t)}\right.$$

$$\left.+\|y(\cdot)\|_{L_p(0,t)}^{p-1}\|w(\cdot)\|_{L_p(0,t)}\right],$$

where

$$M_5 := \mu_{\min}(P)^{-\frac{1}{2}} \cdot M_4,$$

and hence, by (2.6), for all $t \in [0, \omega)$,

$$\frac{1}{p}\|y(t)\|_P^p \leq \frac{1}{p}\|y(0)\|_P^p + M_5 \left[(1 + \frac{M_1\|A_2\|\|A_3\|}{\varepsilon})\|y(\cdot)\|_{L_p(0,t)}^p\right.$$

$$\left. +\|w(\cdot)\|_{L_p(0,\infty)}\|y(\cdot)\|_{L_p(0,t)}^p + \|w(\cdot)\|_{L_p(0,\infty)}\right]$$

$$+ \int_0^t \|y(s)\|^{p-1}\langle \beta(y(s)), PCBu(s)\rangle ds.$$

This proves the inequality. □

We end this section with a useful lemma on minimum phase systems. To this end we first prove a lemma on a property of L_p-functions.

Lemma 2.1.7
Suppose $f(\cdot) : [0,\infty) \to \mathbb{R}^n$ is any absolutely continuous function. If

$$f(\cdot) \in L_p(0,\infty) \quad \text{for some} \quad p \in [1,\infty)$$

and

$$\dot{f}(\cdot) \in L_q(0,\infty) \quad \text{for some} \quad q \in [1,\infty],$$

then

$$f(\cdot) \in L_i(0,\infty) \quad \text{for all} \quad i \in [p,\infty], \quad \text{and} \quad \lim_{t\to\infty} f(t) = 0.$$

Proof: Since $t \mapsto \|f(t)\|$ is a composition of absolutely continuous maps, the set

$$J := \{t \geq 0 \mid \|f(t)\| \text{ is not differentiable}\}$$

is of measure zero. A routine calculation gives

$$\frac{d}{ds}\|f(s)\| = \frac{\langle f(s), \dot{f}(s)\rangle}{\|f(s)\|} \quad \text{for all} \quad s \in \mathbb{R}_+ \setminus J.$$

Put $r = p - \frac{p}{q} + 1$ where $\frac{1}{q} := 0$ if $q = \infty$. Since

$$\int_{[0,\infty)\setminus J} \frac{d}{ds}\|y(s)\|^r ds = \int_{[0,\infty)} \frac{d}{ds}\|y(s)\|^r ds,$$

it follows from Hölder's inequality, that

$$\|f(t)\|^r - \|f(t_n)\|^r = \int_{t_n}^{t} \frac{d}{ds}(\|f(s)\|^r)ds$$

$$= r \int_{t_n}^{t} \|f(s)\|^{r-1} \cdot \frac{d}{ds}(\|f(s)\|)ds$$

$$\leq r \int_{t_n}^{t} \|f(s)\|^{r-1} \cdot \|\dot{f}(s)\|ds$$

$$\leq r \left[\int_{t_n}^{t} \|f(s)\|^{(r-1)\frac{p}{r-1}} \right]^{\frac{r-1}{p}} \cdot \left[\int_{t_n}^{t} \|\dot{f}(s)\|^q ds \right]^{\frac{1}{q}}$$

$$= r \|f(\cdot)\|_{L_p(t_n,t)}^{r-1} \cdot \|\dot{f}(\cdot)\|_{L_q(t_n,t)}$$

for all $t \geq t_n \geq 0$. Therefore,

$$\|f(t)\|^r \leq r\|f(\cdot)\|_{L_p(t_n,\infty)}^{r-1} \cdot \|\dot{f}(\cdot)\|_{L_q(t_n,\infty)} + \|f(t_n)\|^r$$

for all $t \geq t_n$. Choose a sequence $\{t_n\}_{n\in\mathbb{N}}$ so that $\lim_{n\to\infty} t_n = \infty$ and $\lim_{n\to\infty} f(t_n) = 0$. Since

$$\lim_{n\to\infty} \|f(\cdot)\|_{L_p(t_n,\infty)} = \lim_{n\to\infty} \|\dot{f}(\cdot)\|_{L_q(t_n,\infty)} = 0,$$

the above inequality proves $\lim_{t\to\infty} f(t) = 0$. Thus $f(\cdot) \in L_\infty(0,\infty)$ and therefore $f(\cdot) \in L_p(0,\infty) \cap L_\infty(0,\infty)$. This completes the proof.   □

Lemma 2.1.8
Suppose the system

$$\begin{aligned} \dot{x}(t) &= Ax(t) + Bu(t), & x(0) = x_0 \in \mathbb{R}^n \\ y(t) &= Cx(t), \end{aligned} \right\} \quad (2.12)$$

with $(A, B, C) \in \mathbb{R}^{n\times n} \times \mathbb{R}^{n\times m} \times \mathbb{R}^{m\times n}$, is minimum phase and satisfies one of the conditions

(i) $|CB| \neq 0$

(ii) $m = 1, CB = 0, CAB \neq 0, \dfrac{CA^2B}{CAB} < 0.$

If $y(\cdot) \in L_p(0, \infty)$ for some $p \geq 1$, and $\mathcal{K}(\cdot) : \mathbb{R} \to \mathbb{R}^{m \times m}$ is locally integrable and essentially bounded, then the feedback $u(t) = -\mathcal{K}(t)y(t)$ applied to (2.12) yields that the state of the closed-loop system

$$\dot{x}(t) = [A - B\mathcal{K}(t)C]\, x(t), \qquad x_0 \in \mathbb{R}^n$$

satisfies

$$x(\cdot) \in L_p(0, \infty) \cap L_\infty(0, \infty) \qquad \text{and} \qquad \lim_{t \to \infty} x(t) = 0.$$

Proof: If $|CB| \neq 0$, then without loss of generality we may assume that (2.12) is of the form (2.3). Since $\sigma(A_4) \subset \mathbb{C}_-$ and $y(\cdot) \in L_p(0, \infty)$, it follows from the second equation in (2.3) that $z(\cdot) \in L_p(0, \infty)$. Since $\mathcal{K}(\cdot)$ is essentially bounded, it follows again from (2.3) that $\dot{y}(\cdot) \in L_p(0, \infty)$ and $\dot{z}(\cdot) \in L_p(0, \infty)$. Now $x(t) \to 0$ as $t \to \infty$ is a consequence of Lemma 2.1.7. This proves the statement in case of $|CB| \neq 0$.

It remains to prove the lemma if (ii) is valid. Assume, without loss of generality, that (2.12) is of the form (2.4). Since $\sigma(A_6) \subset \mathbb{C}_-$ and $y(\cdot) \in L_p(0, \infty)$, the third equation in (2.4) yields $z(\cdot) \in L_p(0, \infty)$, and hence it follows from the second equation in (2.4), by essential boundedness of $\mathcal{K}(\cdot)$ and $\frac{CA^2B}{CAB} < 0$, that $\dot{y}(\cdot) \in L_p(0, \infty)$. Using (2.4) again yields $\left(y(\cdot), \dot{y}(\cdot), z(\cdot)^T\right)^T$, $\left(\dot{y}(\cdot), y^{(2)}(\cdot), \dot{z}(\cdot)^T\right)^T \in L_p(0, \infty)$ and the claim of the lemma is a consequence of Lemma 2.1.7. $\square$

2.2   High-gain stabilizable systems

Using the results of Section 2.1, we are now in a position to characterize the class of all high-gain stabilizable systems and prove the so called High-Gain Lemma for relative degree 1 systems, and also for relative degree 2 systems.

Definition 2.2.1
A multivariable system associated with the transfer function $G(\cdot) \in \mathbb{R}(s)^{m \times m}$ is called *high-gain stabilizable* if there exist $k^* \in \mathbb{R}$ and $K \in \mathbb{R}^{m \times m}$ such that the closed-loop system

$$y(s) = G(s)u(s), \quad u(s) = -kKy(s) \quad \text{is exponentially stable for all} \quad k \geq k^*,$$

i.e. the transfer function $[I_m + kG(s)K]^{-1}G(s)$ has no poles in $\overline{\mathbb{C}}_+$.

Remark 2.2.2

(i) The zeros of a transfer function $G(s)$ are invariant under output feedback, see e.g. Pugh and Ratcliffe (1981).

(ii) If $G(s)$ is minimum phase, then clearly $G(s)^{-1} \in \mathbb{C}^{m \times m}$ exists for all $s \in \overline{\mathbb{C}}_+$.

(iii) If $G(s)$ is proper with $G(\infty) \in GL_m(\mathbb{R})$, then the number of poles and zeros coincide, see e.g. Maciejowski (1989).

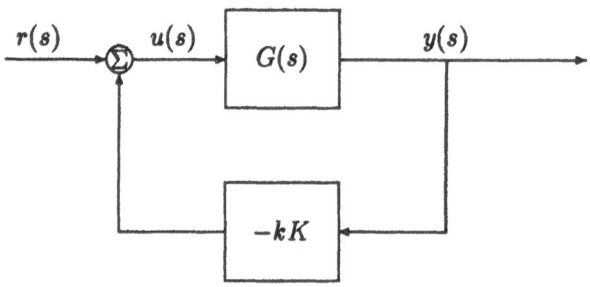

Figure 2.2: Output Feedback Stabilization

Proposition 2.2.3

Let $G(\cdot) \in \mathbb{R}(s)^{m \times m}$ be proper with $\det G(\cdot) \neq 0$, and $K \in \mathbb{R}^{m \times m}$ with $\det K \neq 0$. If feedback of the form $u(s) = -kKy(s)$ is applied to $y(s) = G(s)u(s)$, and k tends to $+\infty$, then as many poles of the closed-loop system as there are zeros are converging to these zeros, the remaining poles tend to infinity.

Proof: Since $\det G(\cdot) \neq 0$, there exists a $k^* > 0$ such that

$$I_m + kG(\cdot)K \in GL_m(\mathbb{R}(s)) \quad \text{for all} \quad k \geq k^*.$$

Therefore, the transfer function of the losed-loop system, for $k \geq k^*$, is given by

$$H_k(s) := [I_m + kG(s)K]^{-1}G(s).$$

Let $\varepsilon(s), \psi(s)$ denote the zeros, poles of $G(s)$, resp., and $\varepsilon(s), \psi_{cl,k}(s)$ denote the zeros, poles of $H_k(s)$, resp. The equality

$$\frac{\varepsilon(s)}{\psi_{cl,k}(s)} = |H_k(s)| = \frac{|G(s)|}{|I_m + kG(s)K|} = \frac{\varepsilon(s)\psi(s)^{-1}}{|I_m + kG(s)K|}$$

yields

$$\psi_{cl,k}(s) = \psi(s)\,|I_m + kG(s)K|.$$

By elementary properties of the determinant we have

$$|I_m + kG(s)K| = 1 + k\alpha_1(s) + \ldots + k^m \alpha_m(s)$$

for some $\alpha_1(\cdot), \ldots, \alpha_{m-1}(\cdot) \in \mathbb{R}(s)$ and $\alpha_m(s) = |G(s)K| = \frac{\varepsilon(s)}{\psi(s)}|K|$. This yields

$$\psi_{cl,k}(s) = \psi(s)[1 + k\alpha_1(s)\ldots + k^{m-1}\alpha_{m-1}(s)] + k^m \varepsilon(s)|K|$$

and the proposition follows from this equality. $\qquad\square$

An immediate consequence of the previous Proposition 2.2.3 is:

Corollary 2.2.4
High-gain stabilizable systems are minimum phase.

For strictly proper minimum phase systems (A, B, C) with $\det(CB) \neq 0$, the statement in Proposition 2.2.3 can be made more explicit in the following sense.

Remark 2.2.5
Suppose (A, B, C) is minimum phase and $K \in \mathbb{R}^{m \times m}$ so that $\sigma(CBK) \subset \mathbb{C}_+$. Using the decomposition (2.3) and $u(t) := -kKy(t)$ together with Schur's formula yields, for $\lambda \notin \sigma(A_1 - kCBK)$,

$$|\lambda I_n - (A - kBKC)| = \begin{vmatrix} \lambda I_m - A_1 + kCBK & -A_2 \\ -A_3 & \lambda I_{n-m} - A_4 \end{vmatrix}$$

$$= |\lambda I_m - A_1 + kCBK| \cdot |(\lambda I_{n-m} - A_4) - A_3(\lambda I_m - A_1 + kCBK)^{-1}A_2| \tag{2.13}$$

or in other words: if k tends to $+\infty$, then the eigenvalues of $A - kBKC$ are approaching the eigenvalues of $-kCBK$ and A_4, where A_4 is a asymptotically stable matrix, so that in the limit we have

$$\lim_{k \to \infty} \sigma(A - kBKC) = \lim_{k \to \infty} \sigma(-kCBK) \cup \sigma(A_4).$$

Thus, there exists a $k' \geq 0$ such that $\sigma(A - kBKC) \subset \mathbb{C}_-$ for all $k \geq k'$. Note that the set $\{k \geq 0 \mid \sigma(A - kBKC) \cap \overline{\mathbb{C}}_+ \neq \emptyset\}$ is not necessarily connected.

For single-input, single-output systems a complete characterization of high-gain (output) stabilizability in terms of the state space matrices can be given as follows.

Theorem 2.2.6
Suppose $(A, b, c, d) \in \mathbb{R}^{n \times n} \times \mathbb{R}^n \times \mathbb{R}^{1 \times n} \times \mathbb{R}$ and $g(\cdot) = c(\cdot I_n - A)^{-1}b + d \not\equiv 0$. $g(\cdot)$ is high-gain stabilizable if, and only if, it is minimum phase and one of the following conditions is satisfied

   (i) $d \neq 0$,

   (ii) $cb \neq 0$,

   (iii) $cAb \neq 0$ and $\frac{cA^2b}{cAb} < 0$.

Proof: It is proved in Proposition 2.2.3 that, under high-gain feedback, $n - r$ poles are tending to the $n - r$ zeros of the system, and the remaining r poles tend to infinity. Moreover, the angles of the asymptotes are $\pm \frac{360}{r}$ deg. See Ogata (1990), p.369. Therefore, high-gain stabilizability implies that (A, b, c, d) is necessarily minimum phase and of relative degree 0, 1, or 2.
If $d \neq 0$, then the number of poles and zeros coincide and the minimum phase assumption yields high-gain stability by virtue of Proposition 2.2.3.
If $cb \neq 0$ and $d = 0$, then it follows from (2.13) and using the notation of Lemma 2.1.3, that under feedback $u(t) = \zeta k y(t)$, where $\zeta \in \{-1, +1\}$, we have, for all $\lambda \notin \sigma(A_1 - \zeta kcb)$,

$$|\lambda I_n - A + \zeta kbc| = (\lambda - A_1 + \zeta kcb) \cdot \left| \lambda I_{n-1} - A_4 - A_3 (\lambda - A_1 + \zeta kcb)^{-1} A_2 \right|$$

and hence

$$\sigma(A - \zeta kbc) \longrightarrow \{A_1 - \zeta kcb\} \cup \sigma(A_4) \quad \text{as} \quad k \to \infty.$$

Since A_4 is asymptotically stable, the proposition is proved for relative degree 1 systems and $\zeta = \frac{cb}{|cb|}$.
If (A, b, c) is of relative degree 2, it follows from (2.4) that, for $\lambda \notin \sigma(A_6)$, we have

$$|\lambda I_n - A + \zeta kbc| = |\lambda I_{n-2} - A_6| \cdot \left| \lambda \left(\lambda - \frac{cA^2b}{cAb} \right) + \zeta k \, cAb - a_2 \right.$$
$$\left. - a_4^T (\lambda I_{n-2} - A_6)^{-1} a_5 \right|$$

If $cA^2b = 0$, then the system is not positive or negative high-gain stabilizable, whereas if $cA^2b \neq 0$, then

$$\sigma(A - \zeta kbc) \longrightarrow \sigma(A_6) \cup \left\{ \frac{cA^2b}{2cAb} \pm \sqrt{\left(\frac{cA^2b}{2cAb} \right)^2 - \zeta k \, cAb} \right\} \quad \text{as} \quad k \to \infty.$$

Choosing $\zeta = \frac{cAb}{|cAb|}$, stability of $A - \zeta kbc$ follows for k sufficiently large. This proves the proposition. □

The following lemma is often called the High-Gain Lemma. It shows that the stability behaviour of the closed-loop system described in Remark 2.2.5 for k large and constant, also holds if $k(t)$ is time-varying.

Lemma 2.2.7
Suppose $A, E \in \mathbb{R}^{n \times n}$, and $k : [0, \infty) \to \mathbb{R}$ is a piecewise continuous function so that

(i) $\sigma(E) \setminus \{0\} \subset \mathbb{C}_+$ and the zero eigenvalues of E are semisimple,
(ii) there exist $\varepsilon, t^* > 0$ such that $\operatorname{Re} \sigma(A - k(t)E) < -\varepsilon$ for all $t \geq t^*$,
(iii) there exists a $k^* > 0$ such that $k(t) \geq k^*$ for all $t \geq t^*$.

If k^* is sufficiently large (in terms of A, E), then there exist $M, \alpha > 0$ such that the solution of the time-varying system

$$\dot{x}(t) = [A - k(t)E]x(t), \qquad x(t_0) = x_0$$

satisfies

$$\|x(t)\| \leq Me^{-\alpha(t - t_0)}\|x_0\| \quad \text{for all} \quad t \geq t_0, \, t_0 \geq 0, \, x_0 \in \mathbb{R}^n,$$

in other words, the system is *exponentially stable*.

Proof: Let $S \in GL_n(\mathbb{R})$ be such that

$$S^{-1}ES = \begin{bmatrix} E_1 & 0 \\ 0 & 0 \end{bmatrix}, \qquad S^{-1}AS = \begin{bmatrix} A_1 & A_2 \\ A_3 & A_4 \end{bmatrix},$$

where $E_1 \in \mathbb{R}^{r \times r}, r = rkE$, and $A_1 \in \mathbb{R}^{r \times r}, A_2, A_3^T \in \mathbb{R}^{r \times (n-r)}$, $A_4 \in \mathbb{R}^{(n-r) \times (n-r)}$ are real matrices. $(\eta^T, z^T)^T := S^{-1}x$ satisfies

$$\frac{d}{dt}\begin{pmatrix} \eta(t) \\ z(t) \end{pmatrix} = \begin{bmatrix} A_1 - k(t)E_1 & A_2 \\ A_3 & A_4 \end{bmatrix}\begin{pmatrix} \eta(t) \\ z(t) \end{pmatrix}. \tag{2.14}$$

We have, by Schur's formula, for all $\lambda \notin \sigma(A_1 - k(t)E_1)$

$$|\lambda I_n - A + k(t)E| = |\lambda I_r - A_1 + k(t)E_1|$$
$$\cdot \left| \lambda I_{n-r} - A_4 - A_3(\lambda I_r - A_1 + k(t)E_1)^{-1}A_2 \right| \tag{2.15}$$

(2.15) yields, by $\sigma(E_1) \subset \mathbb{C}_+$, assumptions (ii), (iii), and for k^* sufficiently large, that $\sigma(A_4) \subset \mathbb{C}_-$.
Define the positive-definite Lyapunov function

$$V(\eta, z) := \frac{1}{2}\langle \eta, R\eta \rangle + \frac{1}{2}\langle z, Qz \rangle$$

where $R = R^T \in \mathbb{R}^{r \times r}$ resp. $Q = Q^T \in \mathbb{R}^{(n-r) \times (n-r)}$ are the positive-definite solutions of

$$RE_1 + E_1^T R = 2I_r, \quad \text{resp.} \quad QA_4 + A_4^T Q = -2I_{n-r}.$$

The derivative of V along the solution of (2.14) is

$$
\begin{aligned}
\frac{d}{dt} V(\eta(t), z(t)) \;=\; & -k(t)\|\eta(t)\|^2 - \|z(t)\|^2 \\
& + \langle \eta(t), R[A_1 \eta(t) + A_2 z(t)] \rangle + \langle z(t), Q A_3 \eta(t) \rangle \\
\leq \;& -(k(t) - M_1)\|\eta(t)\|^2 - \|z(t)\|^2 + \sqrt{2} M_1 \|\eta(t)\| \frac{1}{\sqrt{2}} \|z(t)\| \\
\leq \;& -(k(t) - M_1 - 2M_1^2)\|\eta(t)\|^2 - \frac{1}{2}\|z(t)\|^2
\end{aligned}
$$

where

$$M_1 := \|R A_1\| + \|R A_2\| + \|Q A_3\|.$$

Choosing $k^* > 0$ sufficiently large so that

$$k(t) - M_1 - 2M_1^2 \geq \frac{1}{2} \quad \text{for all} \quad k(t) \geq k^*, \quad t \geq t^*,$$

yields for $\alpha = \frac{1}{2} \min \left\{ \frac{1}{\|R\|}, \frac{1}{\|Q\|} \right\}$ and $t \geq t^*$

$$\frac{d}{dt} V(\eta(t), z(t)) \;\leq\; -\frac{1}{2} \left[\|\eta(t)\|^2 + \|z(t)\|^2 \right] \;\leq\; -2\alpha V(\eta(t), z(t)).$$

Integration of

$$\frac{\frac{d}{ds} V(\eta(s), z(s))}{V(\eta(s), z(s))} \leq -2\alpha$$

over $[t_0, t]$, $t_0 \geq t^*$, and using the fact that

$$\gamma_1 \|(\eta^T, z^T)^T\|^2 \;\leq\; V(\eta, z) \;\leq\; \gamma_2 \|(\eta^T, z^T)^T\|^2$$

where

$$\gamma_1 := \frac{1}{2} \min\{\mu_{\min}(R), \mu_{\min}(Q)\}, \quad \gamma_2 := \frac{1}{2} \left[\|R\| + \|Q\| \right]$$

yields, for all $t \geq t_0 \geq t^*$,

$$
\begin{aligned}
\|x(t)\|^2 \leq \frac{\|S\|^2}{\gamma_1} V(\eta(t), z(t)) \;\leq\; & \frac{\|S\|^2}{\gamma_1} e^{-2\alpha(t - t_0)} V(\eta(t_0), z(t_0)) \\
\leq \;& M_2^2 e^{-2\alpha(t - t_0)} \|x(t_0)\|^2
\end{aligned}
$$

where $M_2 := \sqrt{\gamma_1/\gamma_2}\|S\|$. From this the claim of the lemma follows. $\square$

Remark 2.2.8

(i) Lemma 2.2.7 can be viewed as a *time-varying linear* version of Tychonov's Singular Perturbation Theorem, see e.g. Kokotović (1984). The dynamics are split into fast (here due to $A_1 - kE$, k large) and slow (here due to A_4) dynamics.

(ii) The assumptions (i)-(iii) of Lemma 2.2.7 are satisfied for the class of minimum phase systems $(A, B, C) \in \mathbb{R}^{n \times n} \times \mathbb{R}^{n \times m} \times \mathbb{R}^{m \times n}$ with $\sigma(CB) \subset \mathbb{C}_+$ and $E = BC$, see Lemma 2.1.3. This already indicates how to design an adaptive stabilizer for this special system class: Let $u(t) = -k(t)y(t)$, $\dot{k}(t) = \|y(t)\|^2$, $k(0) = k_0 \in \mathbb{R}$. If the integral $\int_0^t \|y(s)\|^2 ds$ diverges, then Lemma 2.2.7 yields that $x(t)$ is decaying exponentially, thus contradicting that $k(t)$ diverges. Now it needs to be proved that for bounded $k(\cdot)$, and $x(t)$ tends to 0 as $t \to 0$. We will do this in Chapter 4.

A High-Gain Lemma for relative degree 2 systems is more subtle but possible, and will be useful for the proof of stability of adaptive stabilizers considered in Chapter 4.

Lemma 2.2.9

Suppose the system

$$\begin{aligned} \dot{x}(t) &= Ax(t) + bu(t) & x(t_0) = x_0 \\ y(t) &= cx(t) \end{aligned} \right\} \tag{2.16}$$

is minimum phase with

$$cb = 0, \quad cAb \neq 0, \quad \frac{cA^2b}{cAb} < 0, \tag{2.17}$$

and $N(\cdot) : I \to \mathbb{R}$ is a differentiable function on some interval $I := [t_0, t') \subset \mathbb{R}$ with

$$cAbN(t) > M, \quad \text{for all } t \in I \tag{2.18}$$
$$cAb\dot{N}(t) \geq 0, \quad \text{for all } t \in I \tag{2.19}$$

for some $M > 0$.

If feedback of the form $u(t) = -N(t)y(t)$ is applied to (2.16) and M is sufficiently large (depending only on the entries of A, b, c), then there exist $\hat{M}, \hat{\omega} > 0$ (independent of $N(\cdot)$) such that

$$y(t)^2 \leq \hat{M}e^{-\hat{\omega}(t-t_0)}\|x(t_0)\|^2 \quad \text{for all } t \geq t_0, \ t_0, t \in I, \ x(t_0) \in \mathbb{R}^n. \tag{2.20}$$

Proof: By Lemma 2.1.4, we may assume that the closed-loop system (2.16), $u(t) = -N(t)y(t)$, is of the form

$$\frac{d}{dt}\begin{pmatrix} y(t) \\ \dot{y}(t) \\ z(t) \end{pmatrix} = \begin{bmatrix} 0 & 1 & 0 \\ a_2 - cAbN(t), & -a_3 & a_4^T \\ a_5 & 0 & A_6 \end{bmatrix} \begin{pmatrix} y(t) \\ \dot{y}(t) \\ z(t) \end{pmatrix} \qquad (2.21)$$

where

$$a_2 \in \mathbb{R}, \ a_4, \ a_5 \in \mathbb{R}^{n-2}, \ a_3 = \frac{cA^2b}{cAb} > 0, \ \sigma(A_6) \subset \mathbb{C}_-. \qquad (2.22)$$

In order to obtain a Lyapunov-like function for (2.21), we apply the time-varying transformation

$$U(t)^{-1}\begin{pmatrix} y(t) \\ \dot{y}(t) \\ z(t) \end{pmatrix} = \begin{pmatrix} y(t) \\ \eta(t) \\ z(t) \end{pmatrix}, \quad U(t)^{-1} := \begin{bmatrix} 1 & 0 & 0 \\ r(t)\frac{a_3}{2}, & r(t) & 0 \\ 0 & 0 & I_{n-2} \end{bmatrix}$$

for all $t \in I$, where

$$r(t) := \left(cAbN(t) - a_2 - \frac{a_3^2}{4}\right)^{-\frac{1}{2}}$$

and $M > a_2 + \frac{a_3^2}{4}$. Note that it follows from (2.18) that

$$0 < r(t) < \left(cAbM - a_2 - \frac{a_3^2}{4}\right)^{-\frac{1}{2}} \qquad \text{for all} \ \ t \in I. \qquad (2.23)$$

This transformation, due to Corless (1991), has the benefit that the first two off diagonal terms in the transformed system

$$\frac{d}{dt}\begin{pmatrix} y(t) \\ \eta(t) \\ z(t) \end{pmatrix} = \begin{bmatrix} -\frac{a_3}{2} & r(t)^{-1} & 0 \\ -r(t)^{-1}, & -\frac{a_3}{2} + \frac{\dot{r}(t)}{r(t)}, & r(t)a_4^T \\ a_5 & 0 & A_6 \end{bmatrix} \begin{pmatrix} y(t) \\ \eta(t) \\ z(t) \end{pmatrix} \qquad (2.24)$$

are $-r(t)^{-1}$ and $r(t)^{-1}$. Choosing the positive-definite function

$$V(y, \eta, z) := \frac{1}{2}y^2 + \frac{1}{2}\eta^2 + \frac{\alpha}{2}\langle z, Qz \rangle$$

where $Q = Q^T \in \mathbb{R}^{(n-2)\times(n-2)}$ is the positive-definite solution of

$$QA_6 + A_6^T Q = -2I_{n-2}$$

and $\alpha > 0$ is to be specified later, the derivative of V along the solution of (2.24) is

$$\frac{d}{dt}V(y(t), \eta(t), z(t)) = -\frac{a_3}{2}\left[y(t)^2 + \eta(t)^2\right] - \alpha\|z(t)\|^2$$

$$+\frac{\dot{r}(t)}{r(t)}\eta(t)^2 + \eta(t)r(t)a_4^T z(t) + \alpha z(t)^T Qa_5 y(t).$$

Since, by assumption (2.19), $\frac{\dot{r}(t)}{r(t)} = -\frac{1}{2}r(t)^2 cAb\dot{N}(t) < 0$, and

$$\begin{aligned}
\eta(t)r(t)a_4^T z(t) &\leq \tfrac{1}{2}\sqrt{\alpha}\|z(t)\|2\tfrac{1}{\sqrt{\alpha}}r(t)\|a_4^T\|\,|\eta(t)| \\
&\leq \tfrac{1}{4}\alpha\|z(t)\|^2 + \tfrac{4}{\alpha}r(t)^2\|a_4^T\|^2\eta(t)^2 \\
z(t)^T Qa_5 y(t) &\leq \tfrac{1}{2}\|z(t)\|2\|Qa_5\|\,|y(t)| \\
&\leq \tfrac{1}{4}\|z(t)\|^2 + 4\|Qa_5\|^2 y(t)^2,
\end{aligned}$$

we have

$$\begin{aligned}
\frac{d}{dt}V(y(t),\eta(t),z(t)) &\leq \left(-\frac{a_3}{2} + 4\alpha\|Qa_5\|^2\right)y(t)^2 - \frac{1}{2}\alpha\|z(t)\|^2 \\
&\quad + \left(-\frac{a_3}{2} + \frac{4}{\alpha}r(t)^2\|a_4\|^2\right)\eta(t)^2.
\end{aligned}$$

First choose $\alpha > 0$ sufficiently small so that

$$-\frac{a_3}{2} + 4\alpha\|Qa_5\|^2 < -\frac{a_3}{4}$$

and then let M in (2.18) be sufficiently large, see also (2.23), such that

$$-\frac{a_3}{2} + \frac{4}{\alpha}r(t)^2\|a_4\|^2 < -\frac{a_3}{4} \quad \text{for all } t \in I.$$

These assumptions yield for all $t \in I$

$$\frac{d}{dt}V(y(t),\eta(t),z(t)) \leq -\frac{a_3}{4}\left[y(t)^2 + \eta(t)^2\right] - \frac{1}{2}\alpha\|z(t)\|^2 \leq -\hat{\omega}V(y(t),\eta(t),z(t))$$

where $\hat{\omega} = \min\left\{\frac{a_3}{2}, \frac{1}{\|Q\|}\right\}$. Therefore,

$$\frac{1}{2}y(t)^2 \leq V(y(t),\eta(t),z(t)) \leq e^{-\hat{\omega}(t-t_0)}V(y(t_0),\eta(t_0),z(t_0))$$

for all $t \geq t_0$, $t_0, t \in I$. Now it is straightforward to see that

$$V(y(t_0),\eta(t_0),z(t_0)) \leq M_1\,\|(y(t_0),\dot{y}(t_0),z(t_0)^T)^T\|^2$$

where

$$M_1 := \max_{t\in I}\left\{\frac{1}{2} + r(t)^2\left(\frac{a_3^2}{4}+1\right) + \frac{\alpha}{2}\|Q\|\right\}$$

is, by (2.23), independent of $N(\cdot)$. Finally, taking into account the coordinate transformation used in Lemma 2.1.4, we obtain (2.20) and the proof is complete. □

2.3   Notes and References

For strictly proper systems, Proposition 2.1.2 has been proved by Ilchmann and Owens (1992). Lemma 2.1.3 can be found, e.g., in Owens et al. (1984). Lemma 2.1.4 is from Ilchmann and Townley (1992). The input-output representation in (2.5) is due to Logemann and Owens (1988), and becomes useful when considering infinite-dimensional systems. The first inequality in Lemma 2.1.6, in the case of single-input single-output systems, has implicitly been used in Owens et al. (1987) and has been proved in Ilchmann and Owens (1991); for multi-input multi-output systems in the presence of nonlinearities, see Ilchmann and Logemann (1992) for $p = 2$, and for the most general version Ilchmann and Owens (1992). The proof of Lemma 2.1.7 is from Ilchmann and Owens (1992), a different and indirect proof was given independently by Ryan (1992), a proof for p even can be found in Helmke and Prätzel-Wolters (1988).

Proposition 2.2.3 should be known, however I could not find an appropriate reference. The proof is based on techniques used, e.g., in MacFarlane and Postlethwaite (1977). Theorem 2.2.6 is proved in Ilchmann and Townley (1992) for strictly proper systems. Lemma 2.2.7 has independently been claimed by Mårtensson (1986) and Ilchmann et al. (1987), however, Schmid (1991) pointed out that both proof are not correct since the assumption '$\sigma(E) \setminus \{0\} \subset \mathbb{C}_+$' is missing. With this additional assumption, the proof in Ilchmann et al. (1987) goes through. The Lyapunov-function based proof presented here is much simpler. Lemma 2.2.9 is due to Ilchmann and Townley (1992).

For certain classes of **infinite-dimensional** systems the following results are available: An extension of Proposition 2.1.2 to strictly proper multivariable systems is proved in Logemann (1990). The first inequality in Lemma 2.1.6 has been proved implicitly by Logemann and Owens (1988) for single-input, single-output systems and by Logemann and Ilchmann (1992) for multi-input, multi-output systems. The High-Gain Lemma 2.2.7 has been proved by Logemann and Owens (1988) for single-input, single-output systems and for a less general class of infinite-dimensional but multivariable systems by Dahleh (1989).

Chapter 3

Almost Strict Positive Realness

Although strictly positive real systems are a subclass of minimum phase systems, it is worth studying this subclass in depth. It will turn out that minimum phase systems are, modulo an input and output transformation, almost strictly positive real. These transformations are not crucial for universal adaptive stabilization, however the proofs of stability are simpler when using positive realness arguments. This is due to the fact that we are able to use the Lur'e equations. By using this positive real approach, we also will design universal adaptive stabilizers for proper, not necessarily strictly proper, systems.

3.1 Almost strictly positive real systems

In this section the definitions of positive and strictly positive real systems are given. The important analytic and algebraic (in terms of the Lur'e equations) characterizations are stated. This is used to prove that almost strictly positive real systems are high-gain stabilizable and prepares some of the stability proofs for adaptive stabilization in Chapter 4.

Definition 3.1.1
A proper transfer matrix $H(\cdot) \in \mathbb{R}(s)^{m \times m}$ is called *positive real*, if it satisfies

$$\text{all elements of } H(\cdot) \text{ are analytic in } \mathbb{C}_+ \tag{3.1}$$

$$H(s) + H^T(\bar{s}) \geq 0 \quad \text{for all} \quad s \in \mathbb{C}_+. \tag{3.2}$$

$H(s)$ is called *strictly positive real* if there exists an $\varepsilon > 0$ such that $H(s - \varepsilon)$ is positive real.

The next theorem provides an analytic as well as an algebraic characterization of strictly positive real systems.

Theorem 3.1.2
Consider a proper transfer matrix $H(s) = C(sI_n - A)^{-1}B + D$ with minimal realization $(A, B, C, D) \in \mathbb{R}^{n \times n} \times \mathbb{R}^{n \times m} \times \mathbb{R}^{m \times n} \times \mathbb{R}^{m \times m}$, and $\mathrm{rk}\,B = m \leq n$. $H(\cdot)$ is strictly positive real if, and only if, the equivalent statements (i) and (ii) are valid
  (i)

$$\text{all elements of } H(\cdot) \text{ are analytic in } \overline{\mathbb{C}}_+, \tag{3.3}$$

$$H(j\omega) + H^T(-j\omega) > 0 \quad \text{for all} \quad \omega \in \mathbb{R} \tag{3.4}$$

$$\lim_{\omega \to \infty} \omega^2 \left[H(j\omega) + H^T(-j\omega) \right] > 0 \tag{3.5}$$

(ii) there exist $P = P^T \in \mathbb{R}^{n \times n}$ positive-definite, and $Q \in \mathbb{R}^{n \times m}$, $W \in \mathbb{R}^{m \times m}$, $\mu > 0$ so that (A, B, C, D) satisfy the *Lur'e equations*

$$\begin{aligned}
PA + A^T P &= -QQ^T - 2\mu P \tag{3.6} \\
PB &= C^T - QW \tag{3.7} \\
W^T W &= D + D^T. \tag{3.8}
\end{aligned}$$

Proof: see Wen (1988).   □

Remark 3.1.3

  (i) Similar characterizations as in Theorem 3.1.2 have been given by Tao and Ioannou (1988). Their assumptions are different, since they assume that (A, B, C, D) is controllable or observable and that $\mathrm{rk}\left[H(s) + H^T(-s)\right] = m$ for almost all $s \in \mathbb{C}$. If the assumption $\mathrm{rk}\,B = m \leq n$ is not fulfilled, then the input contains redundancies which can be eliminated by a suitable input transformation and a reduced input vector.

 (ii) It immediately follows from (3.7), (3.8) and the controllability of (A, B), that $D = CB = 0$ is impossible. This implies that single-input, single-output, (strictly) positive real systems are either of relative degree 0 or 1.

(iii) If $H(s)$ and $H(\bar{s})$ are invertible, then the relationship

$$H^T(\bar{s})^{-1}\left[H(s) + H^T(\bar{s})\right]H(s)^{-1} = H^T(\bar{s})^{-1} + H(s)^{-1}$$

yields the equivalence

$$H(s) + H^T(\bar{s}) > 0 \quad \Longleftrightarrow \quad H^T(\bar{s})^{-1} + H(s)^{-1} > 0.$$

This is important since it is often easier to prove positive realness of the inverse instead of the nominal system.

A consequence of the next proposition is that positive real systems are high-gain stabilizable.

Proposition 3.1.4
If $G(\cdot) \in \mathbb{R}(s)^{m \times m}$ is strictly positive real and $K \in \mathbb{R}^{m \times m}$ so that $K + K^T > 0$, then the closed-loop system

$$y(s) = G(s)u(s), \qquad u(s) = -kKy(s)$$

is exponentially stable for every $k \geq 0$.

Proof: By definition of strict positive realness it follows that $G(s)^{-1}$ exists for all $s \in \mathbb{C}_+$, and hence

$$|I_m + kG(s)K| = |G(s)| \, |G(s)^{-1} + kK|.$$

By Remark 3.1.3 (iii) and the strict positive realness of $G(s)$, it follows that

$$G(s)^{-1} + kK + (G^T(\bar{s})^{-1} + kK^T) > 0 \quad \text{for all} \quad s \in \mathbb{C}_+,$$

and hence

$$I_m + kG(s)K \in GL_m(\mathbb{C}) \quad \text{for all} \quad s \subset \mathbb{C}_+. \tag{3.9}$$

Let $(A, B, C, D) \in \mathbb{R}^{n \times n} \times \mathbb{R}^{n \times m} \times \mathbb{R}^{m \times n} \times \mathbb{R}^{m \times m}$ denote a minimal realization of $G(\cdot)$ with $\sigma(A) \subset \mathbb{C}_-$. Since $K + K^T > 0$, a similar argument yields

$$I_m + kDK \in GL_m(\mathbb{R}). \tag{3.10}$$

Now (3.10) allows to write the closed-loop system as

$$\begin{aligned}
\dot{x}(t) &= \hat{A}x(t) &&, \hat{A} := A - kBK[I_m + kDK]^{-1}C \\
y(t) &= \hat{C}x(t) &&, \hat{C} := [I_m + kDK]^{-1}C.
\end{aligned}$$

The characteristic polynomial of $\dot{x}(t) = \hat{A}x(t)$ can be expressed as, see MacFarlane and Karcanias (1978) Eqn. (16) or also Kailath (1980) p.651,

$$|sI_n - \hat{A}| = |sI_n - A| \cdot \frac{|I_m + kG(s)K|}{|I_m + kDK|}. \tag{3.11}$$

By (3.9) and (3.10), the fraction in (3.11) is well defined and not zero in $\mathbb{C}_+$. Since A is exponentially stable, the proposition follows from (3.11). □

The notion of strict positive realness will be extended in the sense that we consider the class of systems which are equivalent to strictly positive real systems under output feedback.

Definition 3.1.5
A system $y(s) = G(s)u(s)$, with proper transfer matrix $G(\cdot) \in \mathbb{R}(s)^{m \times m}$, is called *almost strictly positive real*, if there exists a $K \in \mathbb{R}^{m \times m}$ such that the feedback $u(s) = -Ky(s) + r(s)$ yields a strictly positive real system

$$y(s) = H(s)r(s), \qquad H(s) = [I_m + G(s)K]^{-1}G(s).$$

We want to apply feedback of the form $u(t) = -\mathcal{K}(t)y(t)$ to an almost strictly positive real system $y(s) = G(s)u(s)$. In order to make use of the Lur'e equations in Theorem 3.1.2, we first consider a minimal realization of $G(\cdot)$ given by

$$\left. \begin{array}{rcl} \dot{x}(t) & = & Ax(t) + Bu(t) \\ y(t) & = & Cx(t) + Du(t) \end{array} \right\} \tag{3.12}$$

with $(A, B, C, D) \in \mathbb{R}^{n \times n} \times \mathbb{R}^{n \times m} \times \mathbb{R}^{m \times n} \times \mathbb{R}^{m \times m}$. If $K \in \mathbb{R}^{m \times m}$ so that the closed-loop system

$$y(s) = G(s)u(s), \quad u(s) = -Ky(s) + r(s), \quad \text{i.e.} \quad y(s) = H(s)r(s)$$

is strictly positive real, then a minimal state space realization of $y(s) = H(s)r(s)$ is

$$\left. \begin{array}{rcl} \dot{x}(t) & = & \hat{A}x(t) + \hat{B}r(t) \\ y(t) & = & \hat{C}x(t) + \hat{D}r(t) \end{array} \right\} \tag{3.13}$$

where

$$\left. \begin{array}{l} \hat{A} = A - BK[I_m + DK]^{-1}C \qquad\qquad\qquad , \sigma(\hat{A}) \subset \mathbb{C}_- \\ \hat{B} = B - BK[I_m + DK]^{-1}D = B[I_m + KD]^{-1} \\ \hat{C} = C - DK[I_m + DK]^{-1}C = [I_m + KD]^{-1}C \\ \hat{D} = D - DK[I_m + DK]^{-1}D = [I_m + KD]^{-1}D. \end{array} \right\} \tag{3.14}$$

Note that, since the closed-loop transfer matrix is by assumption strictly positive real, and therefore proper rational, it follows from Pugh and Ratcliff (1981) that

$$det[I_m + DK] \neq 0.$$

However if $u(t) = -\mathcal{K}(t)y(t)$ is applied to (3.12), it is not clear, due to the algebraic loop, whether a solution of the closed-loop system exists. We therefore assume that

$$det[I_m + D\mathcal{K}(t)] \neq 0 \quad \text{for all} \quad t \geq 0,$$

and hence the closed-loop system

$$\dot{x}(t) = [A - B\mathcal{K}(t)[I_m + D\mathcal{K}(t)]^{-1}C] \, x(t), \qquad x(0) = x_0$$

is well defined for every piecewise continuous $\mathcal{K}(\cdot) : [0, \infty) \to \mathbb{R}^{m \times m}$.

Using the algebraic characterization of strictly positive real systems in Theorem 3.1.2, the Lur'e equations provide a candidate for a Lyapunov function, so that the behaviour of the solution of the closed-loop system along a positive-definite quadratic function can be described as follows.

Lemma 3.1.6
Suppose $G(\cdot) \in \mathbb{R}(s)^{m \times m}$ is almost strictly positive real, with minimal realization (3.12), and

$$\mathcal{K}(\cdot) : [0, \infty) \to \mathbb{R}^{m \times m} \quad \text{is piecewise continuous,}$$
$$det[I_m + D\mathcal{K}(t)] \neq 0 \quad \text{for all} \quad t \geq 0.$$

Let $K \in \mathbb{R}^{m \times m}$ so that the closed-loop transfer matrix $H(s) = [I_m + G(s)K]^{-1}G(s)$ is strictly positive real.
If $u(t) = -\mathcal{K}(t)y(t)$ is applied to (3.12), then, by Theorem 3.1.2 (ii), there exist $Q \in \mathbb{R}^{n \times m}, W \in \mathbb{R}^{m \times m}, \mu > 0$, and a positive-definite $P = P^T \in \mathbb{R}^{m \times m}$ such that the derivative of the Lyapunov-like function $V(x) = \langle x, Px \rangle$ along the closed-loop system (3.12), $u(t) = -\mathcal{K}(t)y(t)$, which is equivalent to

$$\left. \begin{array}{rcl} \dot{x}(t) & = & \hat{A}x(t) - \hat{B}(\mathcal{K}(t) - K) \, y(t) \\ y(t) & = & \hat{C}x(t) - \hat{D}(\mathcal{K}(t) - K) \, y(t) \end{array} \right\} \qquad (3.15)$$

is given, for all $t \geq 0$, by

$$\begin{array}{rcl} \dfrac{d}{dt}V(x(t)) & = & -2\mu V(x(t)) - \langle y(t), [(\mathcal{K}(t) + \mathcal{K}(t)^T) - (K + K^T)] \, y(t) \rangle \\[2mm] & & -\|Q^Tx(t) - W^T(\mathcal{K}(t) - K) \, y(t)\|^2 \\[2mm] & \leq & -2\mu V(x(t)) - \langle y(t), [\mathcal{K}(t) + \mathcal{K}(t)^T] \, y(t) \rangle + 2\|K\|\|y(t)\|^2. \end{array}$$

Proof: Let $(\hat{A}, \hat{B}, \hat{C}, \hat{D}) \in \mathbb{R}^{n \times n} \times \mathbb{R}^{n \times m} \times \mathbb{R}^{m \times n} \times \mathbb{R}^{m \times m}$ be defined as in (3.14) so that (3.13) denotes a minimal realization of $H(\cdot)$. The decomposition $u(t) = -Ky(t) + r(t)$, where $r(t) = -(\mathcal{K}(t) - K) \, y(t)$, yields that the closed-loop system $u(t) = -\mathcal{K}(t)y(t)$, (3.12), is in fact given by (3.15). It follows from Theorem 3.1.2 that $(\hat{A}, \hat{B}, \hat{C}, \hat{D})$, instead of (A, B, C, D), satisfy (3.6)-(3.8). The derivative of $V(x(t))$ along (3.15) is given by

$$\begin{array}{rcl} \dfrac{d}{dt}V(x(t)) & = & 2\langle x(t), P\hat{A}x(t) \rangle - 2\langle x(t), P\hat{B}(\mathcal{K}(t) - K) \, y(t) \rangle \\[2mm] & = & -\|Q^Tx(t)\|^2 - 2\mu V(x(t)) - 2\langle x(t), \hat{C}^T(\mathcal{K}(t) - K) \, y(t) \rangle \end{array}$$

$$+2\langle x(t), QW^T(\mathcal{K}(t) - K)\,y(t)\rangle$$
$$= \quad -2\mu V(x(t)) - 2\langle y(t) + \hat{D}(\mathcal{K}(t) - K)\,y(t), (\mathcal{K}(t) - K)y(t)\rangle$$
$$-\|Q^T x(t)\|^2 + 2\langle Q^T x(t), W^T(\mathcal{K}(t) - K)\,y(t)\rangle$$
$$= \quad -2\mu V(x(t)) - 2\langle y(t), (\mathcal{K}(t) - K)y(t)\rangle$$
$$-\|Q^T x(t)\|^2 - \langle (\mathcal{K}(t) - K)\,y(t), (\hat{D}^T + \hat{D})(\mathcal{K}(t) - K)y(t)\rangle$$
$$+2\langle Q^T x(t), W^T(\mathcal{K}(t) - K)\,y(t)\rangle.$$

This proves the result. $\square$

Now we are in a position to show that almost strictly positive real systems are high-gain stabilizable, in fact, they have the following stronger property.

Proposition 3.1.7
If $G(\cdot) \in \mathbb{R}(s)^{m \times m}$ is almost strictly positive real, then there exists a $k^* \geq 0$ so that $u(s) = -ky(s)$ yields an exponentially stable closed-loop system

$$H(s) = G(s)[I_m + kG(s)]^{-1} \quad \text{for all} \quad k \geq k^*.$$

Proof: Suppose (3.12) is a minimal realization of $G(\cdot)$ and $K \in \mathbb{R}^{m \times m}$ so that (3.13) is a strictly positive real and minimal realization of $H(\cdot)$. Thus, applying $u(t) = -ky(t)$ to (3.12) is equivalent to applying $r(t) = -[kI_m - K]y(t)$ to (3.13). Since $\hat{D} + \hat{D}^T \geq 0$, it follows that for $k' \geq \|K\|$ we obtain

$$\hat{D}[kI_m - K] + (\hat{D}[kI_m - K])^T \geq 0 \quad \text{for all} \quad k \geq k'.$$

Therefore,
$$I_m + \hat{D}[kI_m - K] \in GL_m(\mathbb{R}) \quad \text{for all} \quad k \geq k'.$$
If $(\hat{A}, \hat{B}, \hat{C}, \hat{D})$ satisfy (3.6)-(3.8), then Lemma 3.1.6 yields, for $V(x) = \langle x, Px\rangle$ and $\mathcal{K}(t) \equiv kI_m$,

$$\frac{d}{dt}V(x(t)) \leq -2\mu V(x(t)) - 2(k - \|K\|)\|y(t)\|^2.$$

Chosing $k \geq \max\{k', \|K\|\}$, $x(t)$ decays exponentially, and hence the proposition follows. $\square$

We like to mention the following implication.

Corollary 3.1.8
If $G(\cdot) \in \mathbb{R}(s)^{m \times m}$ is almost strictly positive real, then it is high-gain stabilizable and therefore, a stabilizable and detectable realization is minimum phase.

3.2 Minimum phase systems

It follows from Proposition 3.1.4 resp. Corollary 3.1.8 that strictly positive resp. almost strictly positive real systems are minimum phase. The converse is not true since a minimum phase system $g(s) = c(sI_n - A)^{-1}b$ does not necessarily satisfy $cb > 0$, which is necessary for positive real systems. However, in this section it will be shown that for large classes of multivariable minimum phase systems, there exist simple input-output transformations, so that the transformed system is almost strictly positive real. More importantly, these transformation are not relevant when constructing universal adaptive stabilizers in Chapter 4, where some proofs of stability are based on positive realness properties.

Proposition 3.2.1
Suppose $G(\cdot) \in \mathbb{R}(s)^{m \times m}$ is minimum phase with minimal realization

$$\begin{aligned} \dot{x}(t) &= Ax(t) + Bu(t) \\ y(t) &= Cx(t) + Du(t) \end{aligned}$$

$(A, B, C, D) \in \mathbb{R}^{n \times n} \times \mathbb{R}^{n \times m} \times \mathbb{R}^{m \times n} \times \mathbb{R}^{m \times m}$ so that $\mathrm{rk}\, B = m \leq n$ and either $det(D) \neq 0$ or, $D = 0$ and $det(CB) \neq 0$. If $R = R^T \in \mathbb{R}^{m \times m}$ is positive-definite, then the closed-loop system

$$y(s) = G(s)U^{-1}\bar{u}(s), \qquad \bar{u}(s) = -ky(s),$$

where

$$\begin{aligned} U &= RCB &&, \text{ if } \quad D = 0 \quad \text{and} \quad det(CB) \neq 0, \\ U &\in GL_m(\mathbb{R}) &&, \text{ if } \quad |D| \neq 0, \end{aligned}$$

is strictly positive real for all $k \geq k^*$, and $k^* > 0$ sufficiently large.

Proof: (a): First consider the case $D = 0$ and $det(CB) \neq 0$. Since the assumptions of Theorem 3.1.2 are satisfied, we shall establish that the transfer matrix of the closed-loop system

$$H_k(s) = \tilde{G}(s)[I_m + k\tilde{G}(s)]^{-1},$$

where

$$\tilde{G}(s) := G(s)U^{-1} = R^{-1}s^{-1} + CABU^{-1}s^{-2} + \dots,$$

satisfies (3.3)-(3.5) for all $k \geq k^*$, k^* is to be determined.

(b): It will be shown that there exist $\omega, k_1 > 0$ such that

$$Re\, \sigma(A - kBU^{-1}C) \leq -\omega \quad \text{for all} \quad k \geq k_1. \tag{3.16}$$

It follows from Lemma 2.1.3 that (A, BU^{-1}, C) is state space equivalent to

$$\left(\begin{bmatrix} A_1 & A_2 \\ A_3 & A_4 \end{bmatrix}, \begin{bmatrix} R^{-1} \\ 0 \end{bmatrix}, [I_m, 0] \right), \quad \text{with} \quad \sigma(A_4) \subset \mathbb{C}_-.$$

Therefore, the feedback system $y(s) = \tilde{G}(s)\bar{u}(s)$, $\bar{u}(t) = -ky(t)$, is equivalent to

$$\begin{aligned} \dot{y}(t) &= [A_1 - kR^{-1}]y(t) &+A_2 z(t) \\ \dot{z}(t) &= A_3 y(t) &+A_4 z(t) \end{aligned} \right\} \tag{3.17}$$

Now Lemma 2.2.7 applied to (3.17) yields (3.16).

(c): Since $\tilde{G}(s)$ is minimum phase, Remark 2.2.2 (i) yields that $H_k(s)$ is minimum phase as well. Hence

$$H_k(s) \in GL_m(\mathbb{C}) \quad \text{for all } s \in \overline{\mathbb{C}}_+ \text{ and all } k \geq k_1 \tag{3.18}$$

Furthermore,

$$H_k(s)^{-1} = [I_m + k\tilde{G}(s)] \, \tilde{G}(s)^{-1} = \tilde{G}(s)^{-1} + kI_m = Rs + kI_m + G_0 + \hat{G}(s)$$

for $G_0 = -RCAB(CB)^{-1}$ and strictly proper $\hat{G}(s)$. Hence there exists a $\gamma > 0$ such that

$$-\gamma I_m < \hat{G}(s) + \hat{G}^T(\bar{s}) + G_0 + G_0^T \quad \text{for all} \quad s \in \overline{\mathbb{C}}_+.$$

Therefore,

$$\begin{aligned} H_k^T(-j\omega)^{-1} + H_k(j\omega)^{-1} &= \hat{G}^T(-j\omega) + \hat{G}^T(j\omega) + G_0 + G_0^T + 2kI_m \\ &> (2k - \gamma)I_m > 0 \end{aligned} \tag{3.19}$$

for all $k > k_2 := \max\{k_1, \frac{\gamma}{2}\}$. By Remark 3.1.3 (iii), it follows from (3.18) and (3.19) that (3.4) holds true for all $k > k_2$.

(d): An elementary calculation gives

$$H_k(s) = R^{-1}s^{-1} - R^{-1}[kI_m + G_0]R^{-1}s^{-2} + \dots$$

and hence, by symmetry of R^{-1},

$$\begin{aligned} &\lim_{\omega \to \infty} \omega^2 \left[H_k(j\omega) + H_k^T(-j\omega) \right] \\ &= \lim_{\omega \to \infty} \left[-\frac{\omega^2}{(j\omega)^2} R^{-1}[kI_m + G_0]R^{-1} - \frac{\omega^2}{(-j\omega)^2}R^{-1}[kI_m + G_0^T]R^{-1} \right] \\ &= 2kR^{-2} + R^{-1}(G_0 + G_0^T)R^{-1}. \end{aligned}$$

This yields (3.5) for all $k \geq k_3$, where $k_3 > k_2$ is chosen so that

$$2kR^{-2} + R^{-1}(G_0 + G_0^T)R^{-1} > 0 \quad \text{for all} \quad k \geq k_3,$$

and completes the first part of the proof.

(e): It remains to consider the case $det(D) \neq 0$. Again, we shall establish (3.3)-(3.5). Since D is invertible, the number of poles coincide with the number of zeros, whence, by the minimum phase assumption and Remark 2.2.2, Proposition 2.2.3 ensures the existence of some $k_1 > 0$ so that the closed-loop transfer matrix

$$H_k(s) = \tilde{G}(s)[I_m + k\tilde{G}(s)]^{-1}, \qquad \tilde{G}(s) = G(s)U^{-1}$$

is exponentially stable for all $k \geq k_1$. This proves (3.3).

(f): Since $H_k(\cdot)$ is stable and minimum phase for all $k \geq k_1$, it follows that

$$H_k(s) \in GL_m(\mathbb{C}) \qquad \text{for all} \quad k \geq k_1 \quad \text{and all} \quad s \in \overline{\mathbb{C}}_+.$$

Using the notation $K = DU^{-1}$, we obtain

$$H_k(s)^{-1} = \tilde{G}(s)^{-1} + kI_m = kI_m + K^{-1} + \hat{G}(s)$$

with $\hat{G}(\cdot)$ strictly proper. Hence there exists a $\gamma > 0$ such that

$$-\gamma I_m < \hat{G}(s) + \hat{G}^T(\bar{s}) \qquad \text{for all} \quad s \in \overline{\mathbb{C}}_+.$$

This yields

$$
\begin{aligned}
H_k^T(-j\omega)^{-1} + H_k(j\omega)^{-1} &= 2kI_m + K^{-T} + K^{-1} + \hat{G}^T(-j\omega) + \hat{G}^T(j\omega) \\
&> 2kI_m + K^{-T} + K^{-1} - \gamma I_m. \qquad (3.20)
\end{aligned}
$$

If $k_2 \geq k_1$ is chosen so that

$$2kI_m + K^{-T} + K^{-1} - \gamma I_m > 0 \qquad \text{for all} \quad k \geq k_2,$$

then it follows from (3.20) together with Remark 3.1.3 (iii) that (3.4) holds true for all $k \geq k_2$.

(g): Since $\lim_{k\to\infty} \left[kI_m + K^{-1}\right]^{-1} = k^{-1}I_m$, there exists a $k_3 \geq k_2$ so that

$$\left[kI_m + K^{-1}\right]^{-1} + \left[kI_m + K^{-1}\right]^{-T} > 0 \quad \text{for all} \quad k \geq k_3.$$

Therefore, $H_k(s)$ can be represented as

$$H_k(s) = \left[kI_m + K^{-1}\right]^{-1} + H_1 s^{-1} + H_2 s^{-2} + \dots$$

for some $H_1, H_2 \in \mathbb{R}^{m \times m}$ and all $k \geq k_3$. Hence,

$$
\begin{aligned}
\lim_{\omega\to\infty} \omega^2 \left[H_k(j\omega) + H_k^T(-j\omega)\right] = \lim_{\omega\to\infty} \Big\{ \omega^2 \left[[kI_m + K^{-1}]^{-1} + [kI_m + K^{-1}]^{-T}\right] \\
+ \frac{\omega}{j}\left[H_1 - H_1^T\right] - \left[H_2 - H_2^T\right] \Big\}
\end{aligned}
$$

$$> 0$$

for all $k \geq k_3$ and hence (3.5) follows for all $k \geq k_3$. This completes the proof of the proposition. $\square$

Corollary 3.2.2
Suppose $(A, b, c, d) \in \mathbb{R}^{n \times n} \times \mathbb{R}^n \times \mathbb{R}^{1 \times n} \times \mathbb{R}$ is minimal, and the transfer matrix

$$g(s) = c(sI_n - A)^{-1}b + d$$

is minimum phase and of relative degree 0 or 1. Then,

$$\tilde{g}(s) := \begin{cases} g(s) & \text{if } d \neq 0 \text{ or } cb > 0 \\ -g(s) & \text{if } d = 0 \text{ and } cb < 0 \end{cases}$$

is almost strictly positive real.

Although the following result covers Proposition 3.2.1, we have chosen to present the less technical proof of the previous proposition separately. Theorem 3.2.3 is not a straightforward generalization of the multivariable case where the spectrum of the high-frequency gain matrix is sign definite. The case '$\det(D) \neq 0$ but D is not of full rank' is more subtle. It is important, since it shows that the multivariable case contains rich classes of systems which can be transformed into an almost strictly positive real system by a suitable input-output transformation. This will be used in Chapter 4 to prove adaptive stabilization results for theses classes.

3.2.3 Theorem[1]
Suppose $G(\cdot) \in \mathbb{R}(s)^{m \times m}$ is minimum phase with minimal realization $(A, B, C, D) \in \mathbb{R}^{n \times n} \times \mathbb{R}^{n \times m} \times \mathbb{R}^{m \times n} \times \mathbb{R}^{m \times m}$ so that $\text{rk}B = m \leq n$. Furthermore, suppose it satisfies (I) and (II):

(I) There exists an orthogonal matrix $[Z, W]$, $Z \in \mathbb{R}^{m \times (m-r)}$, $W \in \mathbb{R}^{m \times r}, r = \text{rk } D$, such that

$$[Z, W]^T D [Z, W] = \begin{bmatrix} 0, & 0 \\ 0, & W^T DW \end{bmatrix} \qquad (3.21)$$

(II) $\det(Z^T CBZ) \neq 0$.

If $R = R^T \in \mathbb{R}^{(m-r) \times (m-r)}$ is positive-definite, then the system

$$\begin{aligned} \dot{x}(t) &= Ax(t) + \bar{B}\bar{u}(t) \\ \bar{y}(t) &= \bar{C}x(t) + \bar{D}\bar{u}(t) \end{aligned} \Bigg\} \qquad (3.22)$$

---

[1] I thank D.H. Owens who pointed out to me a necessary state space condition for positive real systems with D not of full rank. From this I got the idea to state and prove this theorem.

with

$$S := \operatorname{diag}\left\{RZ^TCBZ, I_r\right\}$$
$$\bar{B} = B[Z,W]S^{-1}, \quad \bar{C} = [Z,W]^TC, \quad \bar{D} = [Z,W]^T D[Z,W]S^{-1}$$
$$\bar{u}(t) = S[Z,W]^T u(t), \quad \bar{y}(t) = [Z,W]^T y(t)$$

is almost strictly positive real.

Proof: The assumptions of Theorem 3.1.2 are satisfied and we shall establish (3.3)-(3.5). It follows from the assumptions that

$$\bar{D} = \begin{bmatrix} 0 & 0 \\ 0 & W^TDW \end{bmatrix} \quad \text{and} \quad \bar{C}\bar{B} = \begin{bmatrix} R^{-1} & Z^TCBW \\ W^TCBZ(Z^TCBZ)^{-1}R^{-1}, & W^TCBW \end{bmatrix}.$$

Set

$$B' := BZ \qquad C' := Z^TC.$$

(a): First it is shown that (3.22) is high-gain stabilizable. Let $\bar{u}(t) = -k\bar{y}(t), k \geq 0$. Using the fact that, by invertibility of W^TDW, there exists a $k_1 \geq 0$ so that

$$\left[I_r + kW^TDW\right] \in GL_r(\mathbb{R}) \quad \text{for all} \quad k \geq k_1,$$

a straightforward calculation yields that the closed-loop system is given by

$$\dot{x}(t) = Ax(t) - k(t)\left[B'(Z^TCBZ)^{-1}R^{-1}, BW\right]\bar{y}(t),$$
$$\bar{y}(t) = \begin{bmatrix} C' \\ W^TC \end{bmatrix} x(t) - k\bar{D}\bar{y}(t),$$

respectively

$$\dot{x}(t) = \left[A - kB'(C'B')^{-1}R^{-1}C' - kBW\left[I_r + kW^TDW\right]^{-1}W^TC\right]x(t).$$
$$(3.23)$$

Applying the coordinate transformation

$$U^{-1}x(t) = \begin{bmatrix} R^{-1}C'x(t) \\ Tx(t) \end{bmatrix}, \qquad U := [B'(C'B')^{-1}R, M]$$
$$M \in \mathbb{R}^{n \times (n-m+r)} \quad \text{of full rank so that} \quad \ker C' = M \cdot \mathbb{R}^{n-m+r}$$
$$T := [M^TM]^{-1}M^T\left[I_n - B'(C'B')^{-1}C'\right]$$

to (3.23) yields

$$\frac{d}{dt}\left(U^{-1}x(t)\right) = \left[-k\begin{bmatrix} R^{-1} & 0 \\ 0 & 0 \end{bmatrix} + \tilde{A}_k\right]U^{-1}x(t),$$

where

$$\tilde{A}_k := U^{-1}AU - kU^{-1}BW\left[I_r + kW^TDW\right]^{-1}W^TCU.$$

Since $\tilde{A}_k$ tends to a constant matrix as $k \to \infty$, and since $R^{-1} > 0$, it follows that the poles which do not approach the stable zeros go to $-\infty$. This proves the existence of some $k_2 > k_1$ such that the closed-loop system (3.22), $\bar{u}(t) = -k\bar{y}(t)$, is exponentially stable for all $k \geq k_2$.

(b):   Since $G(\cdot)$ is minimum phase,

$$\bar{G}(s) := [Z, W]^T G(s)[Z, W]S^{-1} = \bar{C}(sI_n - A)^{-1}\bar{B} + \bar{D}$$

is minimum phase as well. There exists a $k_3 \geq k_2$ so that

$$\left[I_m + k\bar{G}(\cdot)\right] \in GL_m(\mathbb{R}(s)) \quad \text{for all} \quad k \geq k_3.$$

Again, by the minimum phase property of $\bar{G}(\cdot)$, its inverse exists and has the form

$$\bar{G}(s)^{-1} = G_p s^p + \ldots + G_1 s + G_0 + \hat{G}(s)$$

for $G_i \in \mathbb{R}^{m \times m}$, $i = 0, \ldots, p$, and strictly proper $\hat{G}(\cdot) \in \mathbb{R}(s)^{m \times m}$. If $p \geq 2$, then comparing the coefficients of $I_m = \bar{G}(s)^{-1}\bar{G}(s) = \bar{G}(s)\bar{G}(s)^{-1}$ at s^p and s^{p-1} yields $G_p = 0$. Therefore,

$$\bar{G}(s)^{-1} = G_1 s + G_0 + \hat{G}(s),$$

and comparing coefficients again gives

$$G_1 = \begin{bmatrix} R & 0 \\ 0 & 0 \end{bmatrix}.$$

Thus the transfer matrix of the closed-loop system

$$H_k(s) := \bar{G}(s)[I_m + k\bar{G}(s)]^{-1}, \qquad k \geq k_3$$

satisfies

$$H_k(s)^{-1} = \begin{bmatrix} R & 0 \\ 0 & 0 \end{bmatrix} s + G_0 + kI_m + \hat{G}(s)$$

with $\hat{G}(\cdot)$ is strictly proper. Since $\hat{G}(\cdot)$ does not have any poles in $\overline{\mathbb{C}}_+$, due to the minimum phase property of $\bar{G}(\cdot)$, there exists a $\gamma > 0$ so that

$$-\gamma I_m < \hat{G}(s) + \hat{G}(s)^T \qquad \text{for all} \quad s \in \overline{\mathbb{C}}_+.$$

If $k_4 \geq k_3$ is defined so that

$$G_0^T + G_0 \geq -(2k - \gamma)I_m \qquad \text{for all} \quad k \geq k_4,$$

then

$$
H_k^T(-j\omega)^{-1} + H_k(j\omega)^{-1} = -j\omega \begin{bmatrix} R & 0 \\ 0 & 0 \end{bmatrix}^T + G_0^T + \hat{G}^T(-j\omega) + 2kI_m
$$

$$
+j\omega \begin{bmatrix} R & 0 \\ 0 & 0 \end{bmatrix} + G_0 + \hat{G}(j\omega)
$$

$$
> -\gamma I_m + (G_0^T + G_0) + 2kI_m
$$

$$
\geq 0.
$$

It follows from Remark 3.1.3 (iii) that (3.4) holds true for all $k \geq k_4$.

(c): It remains to show (3.5). Consider again

$$
\bar{G}(s)^{-1} = \begin{bmatrix} R & 0 \\ 0 & 0 \end{bmatrix} s + G_0 + \hat{G}(s)
$$

with strictly proper $\hat{G}(\cdot) \in \mathbb{R}(s)^{m \times m}$. Comparing coefficients of $I_m = \bar{G}(s)^{-1}\bar{G}(s) = \bar{G}(s)\bar{G}(s)^{-1}$ at s^0 yields

$$
G_0 = \begin{bmatrix} G_0^{11} & * \\ * & (W^T DW)^{-1} \end{bmatrix}.
$$

$H_k(s)$ has the form

$$
H_k(s) = H_p s^p + \ldots + H_1 s + H_0 + H_{-1} s^{-1} + H_{-2} s^{-2} + \ldots
$$

with $H_i \in \mathbb{R}^{m \times m}$, $i = -2, \ldots, p$. Supposing $p > 0$ and comparing coefficients of $I_m = H_k(s)^{-1} H_k(s) = H_k(s)H_k(s)^{-1}$ yields $H_p = 0$. Therefore,

$$
H_k(s) = H_0 + H_{-1} s^{-1} + H_{-2} s^{-2} + \ldots
$$

and

$$
H_k(s)^{-1} = \begin{bmatrix} R & 0 \\ 0 & 0 \end{bmatrix} s + G_0 + kI_m + G_{-1} s^{-1} + \ldots
$$

for some $G_{-1} \in \mathbb{R}^{m \times m}$. We obtain, by comparing coefficients of $I_m = H_k(s)^{-1} H_k(s) = H_k(s)H_k(s)^{-1}$ at s^1, s^0, and s^{-1}

$$
H_k(s) = \begin{bmatrix} 0 & 0 \\ 0 & H_0^{22} \end{bmatrix} + \begin{bmatrix} R^{-1} & * \\ * & * \end{bmatrix} s^{-1} + \begin{bmatrix} H_{-2}^{11} & * \\ * & * \end{bmatrix} s^{-2} + \ldots
$$

where

$$
H_0^{22} = \left[(W^T DW)^{-1} + kI_r \right]^{-1}, \qquad H_{-2}^{11} = -kR^{-2} - R^{-1}G_0^{11}R^{-1}.
$$

Since

$$\lim_{\omega\to\infty} \omega^2 \left\{ H_k(j\omega) + H_k^T(-j\omega) \right\}$$

$$= \lim_{\omega\to\infty} \left\{ \omega^2 \begin{bmatrix} 0 & 0 \\ 0, & H_0^{22} + (H_0^{22})^T \end{bmatrix} + \frac{\omega^2}{j\omega} \begin{bmatrix} 0 & N_1 \\ N_1^T & N_2 \end{bmatrix} \right.$$

$$\left. + \frac{\omega^2}{(j\omega)^2} \begin{bmatrix} H_{-2}^{11} + (H_{-2}^{11})^T, & M_1 \\ M_1 & M_2 \end{bmatrix} \right\}$$

$$= \lim_{\omega\to\infty} \begin{bmatrix} -H_{-2}^{11} - (H_{-2}^{11})^T, & \frac{\omega}{j}N_1 - M_1 \\ -\frac{\omega}{j}N_1^T - M_1, & \omega^2 \left(H_0^{22} + (H_0^{22})^T\right) + \frac{\omega}{j}N_2 - M_2 \end{bmatrix}$$

for some $N_1 = -N_1^T$, $M_1 = M_1^T \in \mathbb{R}^{(m-r)\times(m-r)}$, $N_2 = -N_2^T$, $M_2 = M_2^T \in \mathbb{R}^{r\times r}$ which may depend on k. Now choose $k_5 > k_4$ sufficiently large so that

$$-H_{-2}^{11} - (H_{-2}^{11})^T = 2kR^{-2} + R^{-1}G_0^{11}R^{-1} + (R^{-1}G_0^{11}R^{-1})^T > 0$$

and

$$H_0^{22} + (H_0^{22})^T = \left[(W^T DW)^{-1} + kI_r\right]^{-1} + \left[(W^T DW)^{-1} + kI_r\right]^{-T} > 0$$

for all $k \geq k_5$. It is easily verified that (3.24) yields (3.5).
This completes the proof.   □

We will show that, if K is a stabilizing feedback matrix for a strictly proper system $G(\cdot)$, then the inverse of K implemented as a feedforward loop to $G(\cdot)$ results in an almost strictly positive real system.

Proposition 3.2.4
Suppose $G(\cdot) \in \mathbb{R}(s)^{m\times m}$ is strictly proper with minimal realization $(A, B, C) \in \mathbb{R}^{n\times n} \times \mathbb{R}^{n\times m} \times \mathbb{R}^{m\times n}$ such that $\mathrm{rk}\, B = m \leq n$. If $K \in GL_m(\mathbb{R})$ so that the feedback system

$$y(s) = G(s)u(s), \qquad u(s) = -Ky(s)$$

is exponentially stable, then

$$F(s) := G(s) + K^{-1}$$

is almost strictly positive real.

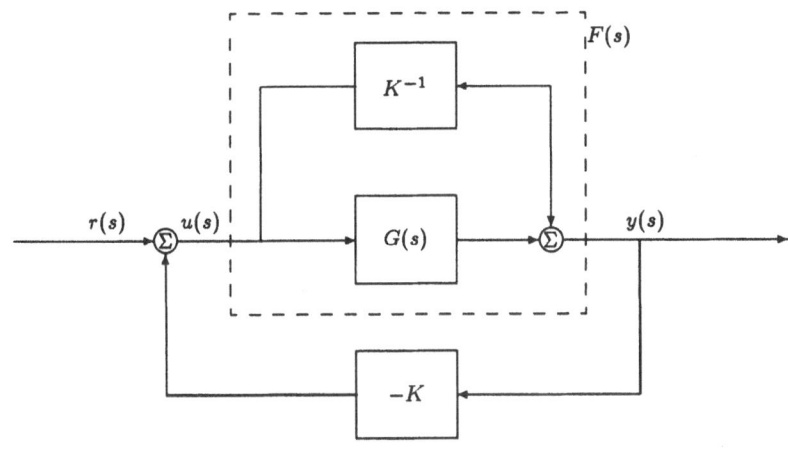

Figure 3.1: Almost strict positive realness via feedforward

Proof: We have to show that there exists a $k^* \geq 0$ such that the transfer matrix

$$H_k(s) := [G(s) + K^{-1}] \cdot [I_m + k(G(s) + K^{-1})]^{-1}$$

is strictly positive real for all $k \geq k^*$. Let $k_1 > 0$ such that

$$I_m + kK^{-1} \in GL_m(\mathbb{R}) \quad \text{for all} \quad k \geq k_1.$$

A minimal state space realization of $F(\cdot)$ is given by

$$\left.\begin{array}{rcl} \dot{x}(t) & = & Ax(t) + Bu(t) \\ \bar{y}(t) & = & Cx(t) + K^{-1}u(t), \end{array}\right\} \tag{3.25}$$

and applying the feedback $y(t) = -ku(t) + r(t)$ to (3.25) yields, for $k \geq k_1$,

$$\begin{array}{rcl} \dot{x}(t) & = & \left[A - kB\left[I_m + kK^{-1}\right]^{-1}C\right]x(t) \quad +B\left[I_m + kK^{-1}\right]^{-1}r(t) \\ \bar{y}(t) & = & \left[I_m + kK^{-1}\right]^{-1}Cx(t) \quad +\left[I_m + kK^{-1}\right]^{-1}K^{-1}r(t), \end{array}$$

which is a minimal state space realization of $H_k(\cdot)$ and, moreover, $\mu_{\min}\left(B[I_m + kK^{-1}]^{-1}\right) > 0$. Since, by assumption, $\sigma(A - BKC) \subset \mathbb{C}_-$, the relationship

$$|F(s)| = \frac{|K^{-1}|}{|sI_n - A|} \cdot |sI_n - [A - BKC]|$$

(see MacFarlane and Karcanias (1978) eqn. (16) or also Kailath (1980) p. 651) yields that $F(\cdot)$ is minimum phase, and hence $H_k(\cdot)$ is minimum phase as well.

Therefore, the assumptions of Theorem 3.2.1 are satisfied and the result follows.   □

Remark 3.2.5

(i) Proposition 3.2.4 provides only the *possibility* to obtain an almost strictly positive real system by augmenting the inverse of a stabilizing feedback gain of a strictly proper system into the feedforward loop. It is not clear how to use it in an adaptive context.

(ii) The same result as in Proposition 3.2.4 can be achieved if the assumption 'rk$B = m \leq n$' is replaced by 'rk$\left[G(s) + G^T(-s)\right] = m$ for almost all $s \in \mathbb{C}$ '. In this case the proof uses the characterization of strictly positive real systems given by Tao and Ioannou (1988).

3.3   Notes and References

The concept of positive real transfer functions and matrices goes back to the 1930s. It is extensively studied in Anderson and Vongpanitlerd (1973) and in Narendra and Taylor (1973), the latter have also introduced the notion of strict positive realness. The concept of *almost* strictly positive real systems is due to BarKana and Kaufman (1985) who used it for model reference adaptive control and strictly proper systems. The idea to concider the Lyapunov-like function $V(x) = \langle x, Px \rangle$ in Lemma 3.1.6 is standard for positive real systems. BarKana and Kaufman (1985) have also claimed Proposition 3.2.4 but without the assumption 'rk$B = m \leq n$', moreover, their proof is not convincing. Some of the results presented in Chapter 3 can be found for discrete time systems in Prätzel-Wolters and Reinke (1991).

Chapter 4

Universal Adaptive Stabilization

In Section 4.1, we shall introduce the concept of switching functions, including Nussbaum-like switching strategies and strategies based on a switching decision function. This, together with the results prepared in Chapter 2 and 3, enables us to present different universal adaptive stabilizers for various classes of systems. The feedback strategy $u(t) = -F(k(t))y(t)$ can be divided into three types, whether $F : \mathbb{R} \to \mathbb{R}^{m \times m}$ is *continuous, piecewise continuous* or *piecewise constant*. In Section 4.2, a continuous or piecewise continuous feedback strategy in combination with $k(t) = \|y(t)\|^p$ is applied to multivariable minimum phase systems (A, B, C, D) with weak assumptions on CB resp. D. In Section 4.3, we introduce a switching strategy which is closer related to the systems dynamics but cannot be applied in the multivariable case if it is only known that $det(CB) \neq 0$. Finally, in Section 4.4 and 4.5, two different modifications are introduced in order to achieve exponential decay of the state $x(t)$. One approach uses an exponential weighting factor in the gain adaptation, while the other one is based on a modification of the feedback strategy introduced in Section 4.2 and uses piecewise constant feedback implementation. For completeness, there should be another section on universal adaptive λ-stabilization. However, this is omitted since it follows immediately from the λ-tracking results in Section 5.2 by setting the class of reference signals $\mathcal{Y}_{ref} = \{0\}$.

4.1   Switching functions

As it was shown in the introduction, the scalar system

$$\left.\begin{array}{rcl} \dot{x}(t) & = & ax(t) + bu(t), \qquad x(0) \in \mathbb{R} \\ y(t) & = & cx(t) \end{array}\right\} \tag{4.1}$$

can be adaptively stabilized by

$$\begin{array}{rcl} u(t) & = & -k(t)y(t) \\ \dot{k}(t) & = & y(t)^2 \qquad\qquad , k(0) \in \mathbb{R} \end{array}$$

provided that $cb > 0$. If the sign of the high-frequency gain is not known, but $cb \neq 0$, then the correct sign of the feedback has to be found adaptively. Applying, for example,

$$\left.\begin{array}{rcl} u(t) & = & -k(t)^2 \cos k(t)\, y(t) \\ \dot{k}(t) & = & y(t)^2 \qquad\qquad , k(0) \in \mathbb{R} \end{array}\right\} \tag{4.2}$$

to (4.1) yields the closed-loop system

$$\begin{array}{rcl} \dot{y}(t) & = & [a - cbk(t)^2 \cos k(t)]y(t) \quad , y(0) = cx(0) \\ \dot{k}(t) & = & y(t)^2 \qquad\qquad\qquad\qquad , k(0) \in \mathbb{R}. \end{array}$$

Since

$$\frac{d}{d\mu}\left(\frac{1}{2}y(\mu)^2\right) = [a - cbk(\mu)^2 \cos k(\mu)]y(\mu)^2 = [a - cbk(\mu)^2 \cos k(\mu)]\dot{k}(\mu),$$

integration over $[0, t]$ and substituting $\tau := k(\mu)$ yields

$$\begin{array}{rcl} \frac{1}{2}\left(y(t)^2 - y(0)^2\right) & = & \displaystyle\int_0^t [a - cbk(\mu)^2 \cos k(\mu)]\dot{k}(\mu)d\mu \\[3mm] & = & \displaystyle\int_{k(0)}^{k(t)} [a - cb\tau^2 \cos \tau]d\tau \\[3mm] & \leq & [k(t) - k(0)]\left[a - \frac{cb}{k(t) - k(0)}\displaystyle\int_{k(0)}^{k(t)} \tau^2 \cos \tau d\tau\right], \end{array}$$

provided $k(t) > k(0)$. It can be shown that

$$\frac{cb}{k(t) - k(0)}\int_{k(0)}^{k(t)} \tau^2 \cos \tau d\tau$$

takes arbitrary large negative and positive values if $k(t)$ tends to $+\infty$. Therefore $k(t)$ is bounded, since otherwise the right hand side of the above inequality takes negative values, contradicting the boundedness from below of the left hand side. Thus, $y(\cdot) \in L_2(0, \infty)$ and the first equation of the closed-loop system yields $\dot{y}(\cdot) \in L_2(0, \infty)$, and hence $y(t) \to 0$. Apart from considering finite escape time, we have shown that (4.2) is a universal adaptive stabilizer for the class of systems (4.1) with $cb \neq 0$.

The underlying idea of this adaptive feedback strategy is, that the input $u(t)$ changes sign as long as $k(t)$ is unbounded, and the sign of the feedback gain $k(t)^2 \cos k(t)$ stays constant for longer and longer periods, until finally the gain $k(t)^2$ is large enough and the interval where the sign is constant is long enough, so that the closed-loop system has enough time to settle down, $x(t)$ goes to zero exponentially and $k(t)$, as the integral of $y(t)^2$, converges to a finite limit and no more switchings occur. Essential in this approach is that the integral of $k(t)^2 \cos k(t)$ has certain properties. This leads to the following definition.

Definition 4.1.1
Let $k' \in \mathbb{R}$. A piecewise right continuous and locally Lipschitz function $N(\cdot) :$ $[k', \infty) \to \mathbb{R}$ is called a *Nussbaum function* if it satisfies

$$\sup_{k>k_0} \frac{1}{k - k_0} \int_{k_0}^{k} N(\tau)\, d\tau = +\infty \text{ and } \inf_{k>k_0} \frac{1}{k - k_0} \int_{k_0}^{k} N(\tau)\, d\tau = -\infty \quad (4.3)$$

for some $k_0 \in (k', \infty)$. A Nussbaum function is called *scaling-invariant* if, for arbitrary $\alpha, \beta > 0$,

$$\tilde{N}(t) := \begin{cases} \alpha N(t) & \text{if } N(t) \geq 0 \\ \beta N(t) & \text{if } N(t) < 0 \end{cases}$$

is a Nussbaum function as well.

It is easy to see that (4.3) holds true for some $k_0 \in (k', \infty)$ if, and only if, it is valid for all $k_0 \in (k', \infty)$.

Examples 4.1.2
The following functions are Nussbaum functions:

$$
\begin{aligned}
N_1(k) &= k^2 \cos k && , k \in \mathbb{R} \\
N_2(k) &= k \cos \sqrt{|k|} && , k \in \mathbb{R} \\
N_3(k) &= \ln k \cos \sqrt{\ln k} && , k > 1
\end{aligned}
$$

$$
N_4(k) = \begin{cases} k & \text{if} & n^2 \leq |k| < (n+1)^2, & n \text{ even} \\ -k & \text{if} & n^2 \leq |k| < (n+1)^2, & n \text{ odd} \end{cases} \quad , k \in \mathbb{R}
$$

$$
N_5(k) = \begin{cases} k & \text{if} & 0 \leq |k| < \tau_0 \\ k & \text{if} & \tau_n \leq |k| < \tau_{n+1}, & n \text{ even} \\ -k & \text{if} & \tau_n \leq |k| < \tau_{n+1}, & n \text{ odd} \end{cases}
$$
$$
\text{with} \quad \tau_0 > 1, \ \tau_{n+1} := \tau_n^2, \ k \in \mathbb{R}
$$

$$
N_6(k) = \cos\left(\tfrac{\pi}{2}k\right) \cdot e^{(k^2)} \qquad\qquad , k \in \mathbb{R}.
$$

See Figure 4.1.

Of course, the cosine in the above examples can be replaced by sine, and similar modifications.

Logemann and Owens (1988) have proved that $N_6(k)$ is scaling-invariant. This property is important if the nomimal system is subjected to certain nonlinear perturbations and/or for some universal adaptive controllers of multivariable systems.

It is easy to see that $N_1(k), N_2(k), N_4(k), N_5(k)$ are in fact Nussbaum functions, whereas to prove the properties (4.3) for $N_3(k)$ is more subtle and a proof is given below. The function $N_3(k)$ has the property that the periods where the sign is kept constant compared to the increase of the gain is larger than for $N_1(k)$ or $N_2(k)$, this will become important for relative degree two systems. Note also that $\lim_{k \to \infty} \frac{d}{dk} N_3(k) = 0$.

Lemma 4.1.3
The function

$$
N(k) : [k_0, \infty) \to \mathbb{R}, \quad k \mapsto \ln k \cos \sqrt{\ln k} \tag{4.4}
$$

is a Nussbaum function for every $k_0 > 1$.

Proof: By differentiating $N(k)$ it is easily seen that $N(\cdot)$ is locally Lipschitz continuous in $[k_0, \infty)$. To establish (4.3) we may assume, without loss of generality, that $k_0 = 1$. The substitution $\ln \sigma = \omega^2$, $\ln k := \rho^2$ yields

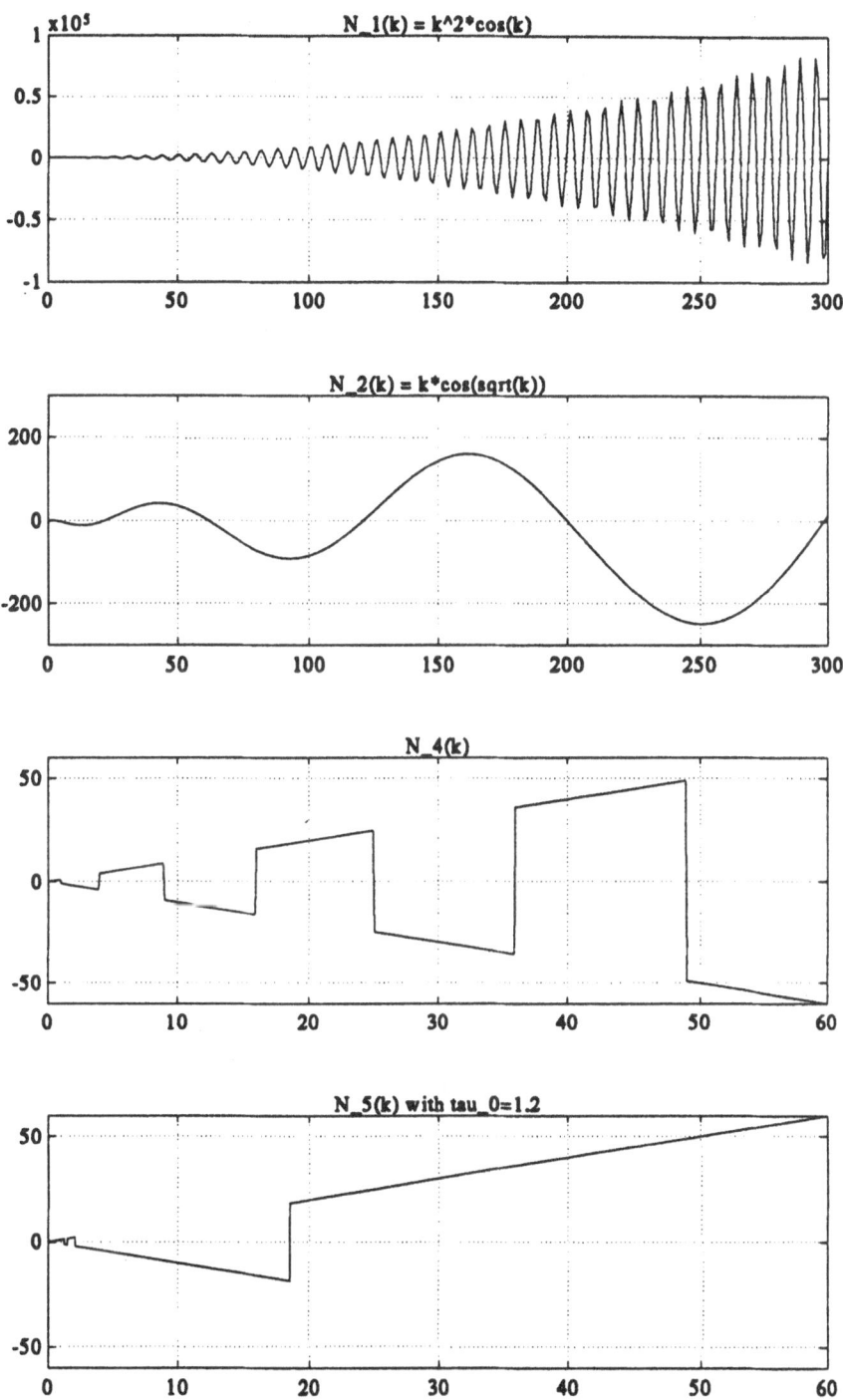

Figure 4.1: Switching functions

$$\int_1^k N(\sigma)\,d\sigma = \int_0^\rho 2\omega^3 e^{\omega^2} \cos\omega\,d\omega$$

$$= e^{\frac{1}{4}}\operatorname{Re}\int_0^\rho 2\omega^3 e^{(\omega+\frac{i}{2})^2}\,d\omega$$

$$= e^{\frac{1}{4}}\operatorname{Re}\left[\int_0^\rho \omega^2 2\left(\omega+\frac{i}{2}\right)e^{(\omega+\frac{i}{2})^2}\,d\omega - i\int_0^\rho \omega^2 e^{(\omega+\frac{i}{2})^2}\,d\omega\right],$$

and integration by parts gives

$$\int_1^k N(\sigma)\,d\sigma = e^{\frac{1}{4}}\operatorname{Re}\left[\left[\omega^2 e^{(\omega+\frac{i}{2})^2}\right]_0^\rho - iI(\rho)\right]$$

where

$$I(\rho) := \int_0^\rho (\omega^2 - 2i\omega)e^{(\omega+\frac{i}{2})^2}\,d\omega$$

$$= \frac{1}{2}\int_0^\rho \omega\,2\left(\omega+\frac{i}{2}\right)e^{(\omega+\frac{i}{2})^2}\,d\omega - \frac{5}{2}\int_0^\rho i$$

$$\left(\omega+\frac{i}{2}\right)e^{(\omega+\frac{i}{2})^2}\,d\omega - \frac{5}{4}\int_0^\rho e^{(\omega+\frac{i}{2})^2}\,d\omega.$$

Again, integration by parts yields

$$I(\rho) = \frac{1}{2}\left[\omega e^{(\omega+\frac{i}{2})^2}\right]_0^\rho - \frac{7}{4}\int_0^\rho e^{(\omega+\frac{i}{2})^2}\,d\omega - \frac{5}{4}i\left[e^{(\omega+\frac{i}{2})^2}\right]_0^\rho.$$

Therefore, we obtain

$$\frac{1}{k-1}\int_1^k N(\sigma)\,d\sigma = \frac{k}{k-1}\left[\left(\ln k - \frac{5}{4}\right)\cos\sqrt{\ln k} + \frac{1}{2}\sqrt{\ln k}\sin\sqrt{\ln k} + \frac{5}{4}\frac{1}{k}\right.$$

$$\left. -\frac{7}{4}\int_0^{\sqrt{\ln k}} e^{-(\ln k - \ln\sigma)}\sin\sqrt{\ln\sigma}\,d\sigma\right]$$

for $k > 1$. This proves the lemma.   □

Nussbaum functions are used if the structural knowledge of the system (A, B, C, D) is: $\sigma(CB)$ lies either in the left- or right-half open plane, for single-input single-output system of relative degree 1 this simply means that the high-frequency gain is nonzero but of unknown sign. In the multivariable case, if it is only known that $det(CB) \neq 0$, the concept needs a generalization. The idea is that the feedback law becomes

$$u(t) = k(t) K_{(Sok)(t)} y(t)$$

where $K_{(Sok)(t)}$ switches between appropriate matrices (in the single-input single-output case simply -1 and $+1$) and, again, the period where the matrix is kept constant must become larger and larger to give the system time to settle down. Due to the following result, $K_{(Sok)(t)}$ has to travel only through a finite set.

Lemma 4.1.4
There exists a finite set

$$\{K_1, \ldots, K_N\} \subset GL_m(\mathbb{R})$$

so that, for any $M \in GL_m(\mathbb{R})$ there exists $i \in \{1, \ldots, N\}$ such that

$$\sigma(M K_i) \subset \mathbb{C}_+$$

Proof: Due to the Polar Decomposition of matrices, $M \in GL_m(\mathbb{R})$ can be factorized as $M = PO$, where $P = \sqrt{M^T M}$ is positive-definite and O is orthogonal,

$$O \in \mathcal{O}_m := \{Q \in GL_m(\mathbb{R}) \,|\, Q^T = Q^{-1}\},$$

see, e.g. Gantmacher (1959) Vol. II, for a proof.
If $\sigma(O) \subset \mathbb{C}_+$, then $\sigma(PO) \subset \mathbb{C}_+$. This follows from the observation that $x \mapsto \langle x, P^{-1}x \rangle$ is a Lyapunov-function for

$$\dot{x}(t) = -P^{-1}Ox(t),$$

since, for all $x(t) \neq 0$,

$$\frac{d}{dt}\langle x(t), P^{-1}x(t) \rangle = -\langle x(t), [O^T + O]x(t) \rangle < 0,$$

where we have used that $\sigma(O) \subset \mathbb{C}_+$ if, and only if, $O + O^T > 0$.
To complete the proof we may therefore assume, without restriction of generality, that M is orthogonal. For every $Q \in \mathcal{O}_m$ there exists $O \in \mathcal{O}_m$ such that $\sigma(QO) \subset \mathbb{C}_+$, simply $O = Q^T$. For fixed O, this property holds also in some

open neighbourhood $\mathcal{B}(O)$. By compactness of $\mathcal{O}_m$, we may choose a finite sub-cover of

$$\mathcal{O}_m = \bigcup_{Q \in \mathcal{O}_m} \mathcal{B}(O).$$

This completes the proof.   □

The set given in Lemma 4.1.4 is often called the *spectrum unmixing set*. Unfortunately, the cardinality of the unmixing sets constructed by Mårtensson (1986,1987) is far too large than would be convenient for applications. Hardly anything is known on the minimum cardinality of unmixing sets, see also Mårtensson (1991). However, for $m = 1$ the set $\{1, -1\}$ is obviously unmixing, while for $m = 2$ there exists an unmixing set of cardinality 6. It has been shown by Zhu (1989) that $GL_3(\mathbb{R})$ can be unmixed by a set having cardinality 32.

While Nussbaum functions can be continuous, we are only able to generalize this concept for piecewise constant functions as for example $N_4(k)$ and $N_5(k)$ in Example 4.1.2.

Definition 4.1.5
Let $N \in \mathbb{N}$. If the sequence $0 < \tau_1 < \tau_2 < \ldots$ satisfies $\lim_{i \to \infty} \tau_i = \infty$, then the associated function

$$S(\cdot) : \mathbb{R} \to \underline{N}, \; k \mapsto S(k) = \begin{cases} 1 & , \text{if } \; k \in (-\infty, \tau_1) \\ i & , \text{if } \; k \in [\tau_{lN+i}, \tau_{lN+i+1}) \end{cases} \quad \begin{array}{l} \text{for some} \\ l \in \mathbb{N}_0, i \in \underline{N} \end{array}$$

is called a *switching function*.

As for Nussbaum functions, the growth of the switching points τ_i is important, for multivariable systems we need the condition

$$\lim_{i \to \infty} \frac{\tau_{i-1}}{\tau_i} = 0. \tag{4.5}$$

Remark 4.1.6
   (i) Obviously, if $\{\tau_i\}_{i \in \mathbb{N}}$ satisfies (4.5), then $\lim_{i \to \infty} \tau_i = \infty$.

   (ii) If $S(\cdot) : \mathbb{R} \to \{1, 2\}$ is a switching function with associated sequence $\{\tau_i\}_{i \in \mathbb{N}}$ satisfying (4.5), and $K_1 = 1, K_2 = -1$ is a spectrum unmixing set for $\mathbb{R} \setminus \{0\}$, then

$$N(k) = k \cdot K_{Sok}$$

is a Nussbaum function.

(iii) Not every piecewise constant Nussbaum function gives rise to a switching function. For example $N_4(k)$ is not a switching function, but $N_5(k)$ is.

(iv) The sequence $\tau_{i+1} := \tau_i + e^{(i^2)}$, $i \in \mathbb{N}$, satisfies (4.5) since

$$\tau_i / \tau_{i+1} = \left[1 + \frac{e^{(i^2)}}{\tau_i} \right]^{-1}$$

and

$$\frac{e^{(i^2)}}{\tau_i} \geq \frac{e^{1+(2i-1)}}{\tau_0 \, e^{-(i-1)^2} + e^{-(i-1)^2} i \, e^{(i-1)^2}} = \frac{e^{1+(2i-1)}}{\tau_0 \, e^{-(i-1)^2} + i} .$$

The right hand side tends to $+\infty$ as i goes to $+\infty$.

Using the concept of switching functions, the Nussbaum functions are generalized as follows.

Lemma 4.1.7
Suppose $S(\cdot) : \mathbb{R} \rightarrow \underline{N}$, $N \in \mathbb{N}$, is a switching function associated with $\{\tau_i\}_{i \in \mathbb{N}}$ satisfying (4.5). Defining, for arbitrary $\alpha > 0$ and every $i \in \underline{N}$,

$$F_i^\alpha(k) := \begin{cases} k & , \text{ if } \quad S(k) = i \\ -\alpha k & , \text{ if } \quad S(k) \neq i \end{cases} \tag{4.6}$$

it follows that $F_i^\alpha(\cdot) : \mathbb{R} \rightarrow \mathbb{R}$ is a scaling-invariant Nussbaum function.

Proof: To prove that

$$\sup_{k>0} \frac{1}{k} \int_0^k F_i^\alpha(x) \, dx = +\infty,$$

we assume, without loss of generality, $i = N$. It follows from the definition of $S(\cdot)$ that

$$\int_{\tau_{jN+1}}^{\tau_{(j+1)N+1}} F_N^\alpha(x) \, dx = - \int_{\tau_{jN+1}}^{\tau_{jN+N}} \alpha \cdot x \, dx + \int_{\tau_{jN+N}}^{\tau_{(j+1)N+1}} x \, dx$$

$$= \frac{\alpha}{2} \tau_{jN+1}^2 - \frac{\alpha+1}{2} \tau_{jN+N}^2 + \frac{1}{2} \tau_{(j+1)N+1}^2 .$$

Therefore,

$$\frac{1}{\tau_{(l+1)N+1} - \tau_1} \int\limits_{\tau_1}^{\tau_{(l+1)N+1}} F_N^\alpha(x)\, dx$$

$$= \frac{1}{\tau_{(l+1)N+1} - \tau_1} \sum_{j=0}^{l} \int\limits_{\tau_{jN+1}}^{\tau_{(j+1)N+1}} F_N^\alpha(x)\, dx$$

$$= \frac{1}{2} \frac{1}{\tau_{(l+1)N+1} - \tau_1} \sum_{j=0}^{l-1} \alpha\, \tau_{jN+1}^2 - [\alpha + 1]\, \tau_{(j+1)N}^2 + \tau_{(j+1)N+1}^2 \qquad (4.7)$$

$$+ \frac{1}{2} \frac{1}{\tau_{(l+1)N+1} - \tau_1} \left[\alpha \cdot \tau_{lN+1}^2 - (\alpha + 1)\tau_{(l+1)N}^2 + \tau_{(l+1)N+1}^2 \right].$$

Since

$$\alpha\, \tau_{jN+1}^2 - [\alpha + 1]\, \tau_{(j+1)N}^2 + \tau_{(j+1)N+1}^2 \geq \tau_{(j+1)N+1}^2 \left[-(\alpha + 1)\frac{\tau_{(j+1)N}^2}{\tau_{(j+1)N+1}^2} + 1 \right],$$

it follows from condition (4.5) that the term in (4.7) is bounded from below by some $L \in \mathbb{R}$ (independent of l). Therefore

$$\frac{1}{\tau_{(l+1)N+1} - \tau_1} \int\limits_{\tau_1}^{\tau_{(l+1)N+1}} F_N^\alpha(x)\, dx \geq L + \frac{1}{2}\frac{\tau_{(l+1)N+1}^2}{\tau_{(l+1)N+1} - \tau_1} \left[1 - (\alpha + 1)\frac{\tau_{(l+1)N}^2}{\tau_{(l+1)N+1}^2} \right],$$

and, again by (4.5), the second summand goes to $+\infty$ as l tends to ∞. It remains to prove

$$\inf_{k>0} \frac{1}{k} \int\limits_0^k F_i^\alpha(x)\, dx \quad = \quad -\infty.$$

Assuming, without loss of generality, $i = 1$, it follows that

$$\int\limits_{\tau_{jN+1}}^{\tau_{(j+1)N+1}} F_N^\alpha(x)\, dx \quad = \quad \frac{1}{2}\left[-\alpha\tau_{(j+1)N+1}^2 + (\alpha + 1)\tau_{jN+2}^2 - \tau_{jN+1}^2 \right]$$

$$\leq \quad -\frac{\alpha}{2}\tau_{(j+1)N+1}^2 \left[1 - \frac{\alpha + 1}{\alpha}\frac{\tau_{jN+2}^2}{\tau_{(j+1)N+1}^2} \right].$$

Again by (4.5), there exists a $L' > 0$ such that

$$\frac{1}{\tau_{(l+1)N+1} - \tau_1} \int_{\tau_1}^{\tau_{(l+1)N+1}} F_N^\alpha(x)\,dx \le L' - \frac{\alpha}{2}\frac{\tau_{(l+1)N+1}^2}{\tau_{(l+1)N+1} - \tau_1}\left[1 - \frac{\alpha+1}{\alpha}\frac{\tau_{(l+1)N}^2}{\tau_{(l+1)N+1}^2}\right].$$

Since the second summand goes to $-\infty$ as l tends to ∞. This proves that $F_i^\alpha(\cdot)$ is a Nussbaum function.
For arbitrary $\alpha', \beta' > 0$ define

$$\tilde{N}(k) := \begin{cases} \alpha' F_i^\alpha(k) &,\text{ if } F_i^\alpha(k) \ge 0 \\ \beta' F_i^\alpha(k) &,\text{ if } F_i^\alpha(k) < 0. \end{cases}$$

For $\gamma := \frac{\alpha\beta'}{\alpha'}$, we have $\tilde{N}(k) = \alpha' F_i^\gamma(k)$ is a Nussbaum function as well, and hence $N(\cdot)$ is scaling-invariant. This completes the proof. $\square$

A different approach to the Nussbaum switching strategy is via a switching decision function which is defined as follows. Suppose $0 < \lambda_1 < \lambda_2 < \ldots$ is a strictly increasing sequence with $\lim_{i\to\infty}\lambda_i = \infty$. Let $k(\cdot) : [0,\infty) \to \mathbb{R}$ with $k(0) = k_0 \in \mathbb{R}$ be a non-decreasing integrable function and $y(\cdot) : [0,\infty) \to \mathbb{R}^m$ be the output of a closed-loop system with feedback

$$u(t) = -k(t)\Theta(t)y(t),$$

where

$$\Theta(\cdot) : [0,\infty) \to \{-1,+1\}$$

is defined via the *switching decision function*

$$\psi(t) = \frac{k_0 + \int_0^t \Theta(\tau)k(\tau)\|y(\tau)\|^p\,d\tau}{1 + \int_0^t \|y(\tau)\|^p\,d\tau}$$

and the algorithm

$$(*)\quad \left.\begin{array}{rcl} i & := & 0 \\ \Theta(0) & := & -1 \quad t(0) := 0 \\ t_{i+1} & := & \inf\{t > t_i\,|\,|\psi(t)| \le \lambda_{i+1}k(0)\} \\ \Theta(t) & := & \Theta(t_i) \quad\text{for all}\quad t \in [t_i, t_{i+1}) \\ \Theta(t_{i+1}) & := & -\Theta(t_i) \\ i & := & i+1 \\ \multicolumn{3}{l}{\text{go to } (*)} \end{array}\right\} \quad (4.8)$$

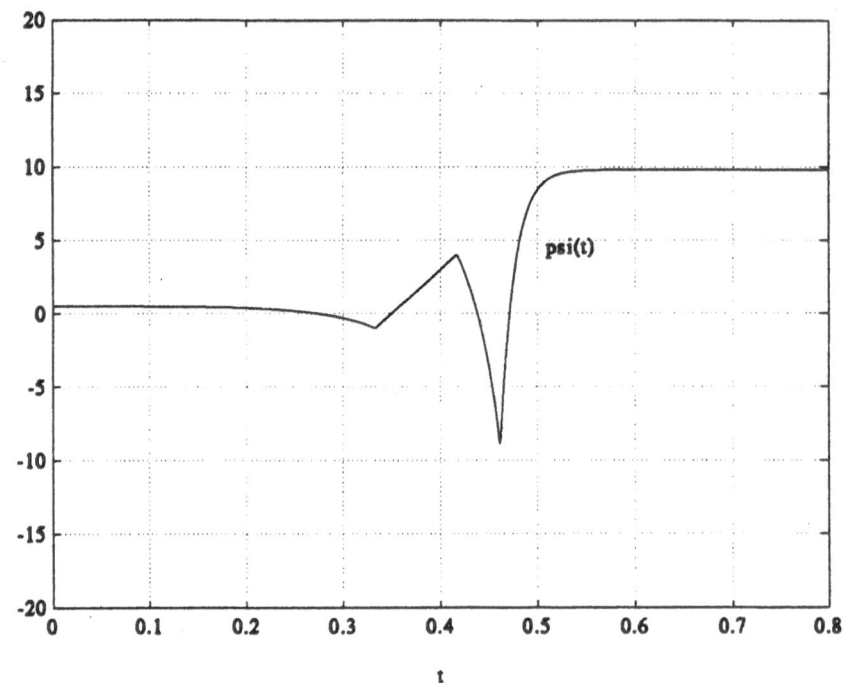

Figure 4.2: Evolution of $\psi(t)$

Remark 4.1.8

(i) The switching function $\Theta(\cdot)$ switches at each time t_i when the switching decision function, which is a stability indicator, reaches the 'threshold' $\lambda_{i+1}k(t_0)$.

(ii) For $k(t) \geq |k_0|$, it is easily calculated that, for every $t \neq t_i$,

$$\frac{d}{dt}\psi(t) = \begin{cases} \geq 0 & , \text{if } \Theta(t) = +1 \\ \leq 0 & , \text{if } \Theta(t) = -1. \end{cases}$$

Moreover, it can be shown that, if $k(t)$ is strictly increasing, then $\psi(t)$ is either strictly increasing or decreasing, taking larger negative and positive values. Therefore, putting $k(t) = \|y(t)\|^p$, the gain $k(t)$ will increase by construction of the switching algorithm (4.8), until finally it will be so large, that the system will be stabilized.

(iii) In Figure 4.2, $k_0 = 0.5$ and $\lambda_1 = 2$, $\lambda_2 = 8$, $\lambda_3 = 16$. In between the switching times the switching decision function $\psi(t)$ is monotonically increasing or decreasing. The simulation belongs to Example 7.1.5 and the system (7.1).

(iv) Note how closely the concept of switching decision functions is related to that of Nussbaum functions. Suppose $k(t) = \|y(t)\|^p > 0$ almost everywhere, then $k^{-1}(\tau) = s$ is well defined and Lemma 4.1.9 below shows that $(\Theta \circ k^{-1})(x) \cdot x$ is a Nussbaum function.

(v) The 1 in the denominator of $\psi(t)$ can be replaced by any positive constant, it only ensures that $\psi(t)$ is well defined if $y(\cdot) \equiv 0$ on some interval $[0, t^*]$.

The advantage of this strategy, which is different to the Nussbaum-type switching strategy, is that the 'stability indicator' $\psi(t)$ is stronger related to the systems dynamics and the controller tolerates large classes of nonlinear disturbances, see Section 6.2. Note also that no assumption is made on how fast the sequence $\{\lambda_i\}_{i \in \mathbb{N}}$ is tending to ∞.

Lemma 4.1.9
Let $p \geq 1$, $\omega \in (0, \infty]$. Suppose $k(\cdot) : [0, \omega) \to \mathbb{R}$ is a piecewise continuous non-decreasing function with $\lim_{t \to \omega} k(t) = \infty$. If $y(\cdot) \notin L_p(0, \omega)$, then the switching decision function $\psi(\cdot) \notin L_\infty(0, \omega)$ and hence, $\psi(t)$ takes arbitrary large negative and positive values.

Proof: Let $k_0 = k(0)$. Suppose $\psi(t)$ is bounded. Thus there exists a $t^* > 0$ such that $\Theta(t) = \Theta(t^*)$ for all $t \geq t^*$. It follows that, for all $t' \in [t^*, \omega)$, we have

$$|\psi(t)| = \left| \frac{k_0 + \int_0^t \Theta(s)k(s)\|y(s)\|^p \, ds}{1 + \int_0^t \|y(s)\|^p \, ds} \right|$$

$$= \left| \frac{k_0 + \int_0^{t'} \Theta(s)k(s)\|y(s)\|^p \, ds}{1 + \int_0^t \|y(s)\|^p \, ds} \right|$$

$$+ |\Theta(t')| \cdot \left| \frac{\int_{t'}^t k(s)\|y(s)\|^p \, ds}{1 + \int_0^{t'} \|y(s)\|^p \, ds + \int_{t'}^t \|y(s)\|^p \, ds} \right|$$

$$\geq \left| \frac{k_0 + \int\limits_0^{t'} \Theta(s)k(s)\|y(s)\|^p \, ds}{1 + \int\limits_0^t \|y(s)\|^p \, ds} \right|$$

$$+ |k(t')| \cdot \frac{\int\limits_{t'}^t \|y(s)\|^p \, ds}{1 + \int\limits_0^{t'} \|y(s)\|^p \, ds + \int\limits_{t'}^t \|y(s)\|^p \, ds}.$$

If $y(\cdot) \notin L_p(0,\omega)$, then $\lim_{t \to \omega} |\psi(t)| \geq |k(t')|$. Hence unboundedness of $k(\cdot)$ yields that $\psi(\cdot)$ is unbounded. This completes the proof.   □

4.2 Stabilization via Nussbaum-type switching

Now we are in a position to put together various results of previous sections (High-Gain Lemmata, inequalities relating present output to past input and output data, relationships between almost strictly positive real and minimum phase systems, switching and Nussbaum functions) in order to present universal adaptive stabilizers for various classes of systems. First we shall show how the simple Willems-Byrnes controller $u(t) = -k(t)y(t)$, $\dot{k}(t) = \|y(t)\|^2$, is extended, via switching functions, to stabilize all strictly proper minimum phase system (A, B, C) with $det(CB) \neq 0$. For single-input, single-output systems a continuous feedback stabilizer is presented in Theorem 4.2.3 which is universal for all high (negative or positive) gain stabilizable systems, therefore including systems of relative degree 2. Apart from the proof of Theorem 4.2.3, in this section all proofs are either based on the inequalities in Lemma 2.1.6 (if strictly proper minimum phase systems are considered) or the inequality in Lemma 3.1.6 which is due to the Lur'e equations (if proper, but not necessarily strictly proper, systems are considered). For a better intuitive understanding of the adaptation mechanisms it may be helpful to consider also the simulations presented in Section 7.1.

In the following theorem it will be shown how the simple Willems-Byrnes controller $\dot{k}(t) = \|y(t)\|^2$, $u(t) = -k(t)y(t)$ can be extended for large classes of systems. We also use the gain adaptation $\dot{k}(t) = \|y(t)\|^p$ for arbitrary $p \geq 1$. The advantage of large p is a better transient behaviour of the output and gain, see Section 7.1.

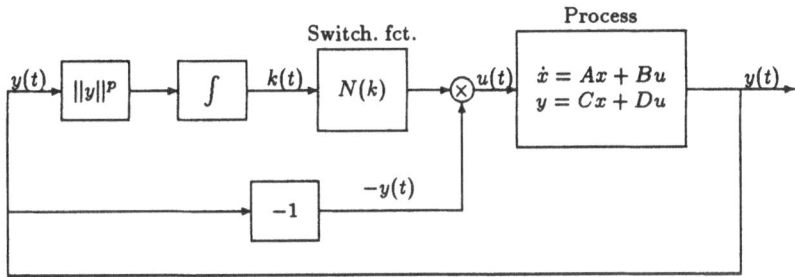

Figure 4.3: Universal adaptive stabilizer via switching function

Theorem 4.2.1

Suppose the system

$$\begin{aligned}
\dot{x}(t) &= Ax(t) + Bu(t) &, x(0) = x_0 \\
y(t) &= Cx(t)
\end{aligned} \right\} \tag{4.9}$$

where $(A, B, C) \in \mathbb{R}^{n \times n} \times \mathbb{R}^{n \times m} \times \mathbb{R}^{m \times n}$, is minimum phase. Let $p \geq 1$, and

$N(\cdot) : \mathbb{R} \to \mathbb{R}$ be a Nussbaum function, scaling-invariant if $m > 1$ or $p \neq 2$,
$\{K_1, \ldots, K_N\}$ a spectrum unmixing set for $GL_m(\mathbb{R})$,
$S(\cdot) : \mathbb{R} \to \underline{N}$ a switching function associated with $\{\tau_i\}_{i \in \mathbb{N}}$ satisfying (4.5).

If the adaptation law

$$\dot{k}(t) = \|y(t)\|^p, \qquad k(0) = k_0$$

together with one of the feedback laws

$(\alpha): \quad u(t) = -k(t)y(t) \qquad\qquad , \text{ if } \sigma(CB) \subset \mathbb{C}_+$

$(\beta): \quad u(t) = -N(k(t))y(t) \qquad , \text{ if } \sigma(CB) \subset \mathbb{C}_+ \text{ or } \mathbb{C}_-$

$(\gamma): \quad u(t) = -k(t)K_{(Sok)(t)}y(t) \quad , \text{ if } |CB| \neq 0$

and arbitrary $k_0 \in \mathbb{R}$, $x_0 \in \mathbb{R}^n$, is applied to (4.9), then the closed-loop system has the properties

(i) the unique solution $(x(\cdot), k(\cdot)) : [0, \infty) \to \mathbb{R}^{n+1}$ exists,

(ii) $\lim_{t \to \infty} k(t) = k_\infty$ exists and is finite,

(iii) $x(\cdot) \in L_p(0,\infty) \cap L_\infty(0,\infty)$ and $\lim_{t\to\infty} x(t) = 0$.

(iv) If (γ) is applied, then there exist $i \in \underline{N}$ and $t^* > 0$, so that $K_{S(k(t))} = K_i$ for all $t \geq t^*$.

The intuition for the adaptive strategy $\dot{k}(t) = \|y(t)\|^2$ and the feedback (α) is as follows: It is known from Remark 2.2.5 that for sufficiently large $k^* > 0$, the feedback system $\dot{x}(t) = [A - k^* BC]x(t)$ is exponentially stable. In the adaptive situation we have $\dot{x}(t) = [A - k(t)BC]x(t)$, and $k(t) = \int_0^t \|y(s)\|^2 ds + k_0$ is monotonically growing. If $x(t)$ is unstable, then $k(t)$ becomes so large that, finally, $x(t)$ decays exponentially (see the High-Gain Lemma 2.2.7), and therefore $k(t)$ tends to a finite limit.

For the intuition behind the feedback strategies (β) and (γ) see Section 4.1.

Proof of Theorem 4.2.1:

(a):   First, existence and uniqueness of the solution of the initial value problem $(\alpha) - (\gamma)$ will be established.

If (γ) is applied, then the closed-loop system is given by

$$\begin{aligned} \dot{x}(t) &= \left[A - k(t)BK_{S(k(t))}C\right] x(t) &, x(0) = x_0 \\ \dot{k}(t) &= \|Cx(t)\|^p &, k(0) = k_0. \end{aligned} \left.\right\} \qquad (4.10)$$

The right hand side is continuous in x, but since $S(\cdot)$ is only piecewise right continuous, it is discontinuous in k. However, since $S(\cdot)$ is piecewise right constant, the right hand side of (4.10) is at least right Lipschitz in x, k at $t = 0$. The same holds true if the feedback (β) is applied, in case of (α), the right hand side is even continuous in k. Therefore, it follows from the classical theory of ordinary differential equations, that for every $(x_0, k_0) \in \mathbb{R}^{n+1}$, the closed-loop system has a unique solution $(x(\cdot), k(\cdot)) : [0, \omega) \to \mathbb{R}^{n+1}$, maximally extended over $[0, \omega)$ for some $\omega > 0$.

It will be proved in steps (b)-(d) that the function $k(\cdot)$ is bounded on its interval of existence.

(b):   We shall establish $k(\cdot) \in L_\infty(0, \omega)$ by contradiction. Suppose $k(\cdot) \notin L_\infty(0, \omega)$ and (α) is applied. Without loss of generality, let (A, B, C) be of the form (2.3). If $\omega = \infty$, then $k(t)$ and $E := \begin{bmatrix} CB & 0 \\ 0 & 0 \end{bmatrix}$ satisfy the assumptions of the High-Gain Lemma 2.2.7. Thus $x(t)$ decays exponentially, and hence $k(t) = \int_0^t \|Cx(s)\|^p ds + k_0$ is bounded. If $\omega \neq \infty$, consider

$$\bar{k}(t) := \begin{cases} k(t) &, \text{ if } t \in [0, \omega - \varepsilon) \\ 0 &, \text{ if } t \geq \omega - \varepsilon \end{cases}$$

for $\varepsilon \in (0, \omega)$. If ε is sufficiently small, then $\bar{k}(t)$ and E satisfy the assumptions of Lemma 2.2.7 again and it follows that $k(\cdot) \in L_\infty(0, \omega)$, contradicting the assumption.

(c): In case of (β), inequality (2.7) yields, for some $M > 0$ and for all $t \in [0, \omega)$,

$$\frac{1}{p}\|y(t)\|_P^p \leq M + M \int_0^t \|y(s)\|^p \, ds - \int_0^t N(k(s))\|y(s)\|_P^{p-1} \langle \beta(y(s)), P\,CBy(s)\rangle ds,$$

where $P = P^T \in \mathbb{R}^{m \times m}$ is the positive-definite solution of

$$PCB + (CB)^T P = 2\sigma I_m$$

for some $\sigma \in \{-1, +1\}$. Using this Lyapunov equation, the adaptation law, and changing variables, we obtain from the inequality above, for $t \in [0, \omega)$ so that $k(t) > k(0)$,

$$
\begin{aligned}
\frac{1}{p}\|y(t)\|_P^p &\leq M + M[k(t) - k(0)] - \int_0^t N(k(s))\sigma\|\beta(y(s))\|\|y(s)\|_P^{p-1}\|y(s)\| ds \\
&\leq M + M[k(t) - k(0)] - \int_0^t \tilde{N}(k(s))\|y(s)\|^p \, ds \\
&\leq M + [k(t) - k(0)]\left[M + \frac{1}{k(t)-k(0)}\int_{k(0)}^{k(t)} \tilde{N}(\tau)d\tau\right]
\end{aligned}
$$

where

$$\tilde{N}(k) := \begin{cases} \sigma\|P\|^{\frac{p-2}{2}} N(k) &, \quad \text{if } \sigma N(k) < 0 \\ \sigma\mu_{\min}(P)^{\frac{p-2}{2}} N(k) &, \quad \text{if } \sigma N(k) \geq 0. \end{cases}$$

In the special case $m = 1$ and $p = 2$, we choose $P = 1$ and derive the same inequality for $\tilde{N}(k) = \sigma N(k)$ by using the first inequality in Lemma 2.1.6. In both cases it follows that $\tilde{N}(\cdot)$ is a switching function as well. If $\lim_{t \to \omega} k(t) = \infty$, then the right hand side of the above inequality takes negative values as $t \to \omega$, thus contradicting the non-negativeness of the left hand side, and hence $k(\cdot) \in L_\infty(0, \omega)$.

(d): If (γ) is applied, choose some $K_i \in \{K_1, \ldots, K_N\}$ so that $\sigma(CBK_i) \subset \mathbb{C}_+$. Let $P = P^T \in \mathbb{R}^{m \times m}$ be the positive-definite solution of

$$PCBK_i + (CBK_i)^T P = 2I_m,$$

and $\alpha' > 0$ so that

$$\|PCBK_l\| \leq \alpha' \quad \text{for all } l \in \underline{N},$$

and define

$$\gamma := \mu_{\min}(P)^{\frac{p-2}{2}}, \qquad \alpha := \gamma^{-1}\alpha'\|P\|^{\frac{p-2}{2}}.$$

Using this notation and $F_i^\alpha(\cdot)$ as defined in (4.6), it follows that, for some $M > 0$ and for all $t \in (0, \omega)$ with $k(t) > k(0)$,

$$-\|y(t)\|_P^{p-1}\langle\beta(y(t)), PCBu(t)\rangle = -k(t)\|y(t)\|_P^{p-1}\langle\beta(y(t)), PCBK_{S(k(t))}y(t)\rangle$$

$$\leq \begin{cases} -k(t)\gamma\|y(t)\|^p & , \text{ if } S(k(t)) = i \\ k(t)\alpha'\|P\|^{\frac{p-2}{2}}\|y(t)\|^p & , \text{ if } S(k(t)) \neq i \end{cases}$$

$$= -\gamma F_i^\alpha(k(t))\dot{k}(t).$$

Inserting this inequality into the one in Lemma 2.1.6 and changing variables yields

$$\frac{1}{p}\|y(t)\|_P^p \leq M + M[k(t) - k(0)] - \gamma\int_0^t F_i^\alpha(k(s))\dot{k}(s)ds$$

$$= M + [k(t) - k(0)]\left[M - \frac{\gamma}{k(t) - k(0)}\int_{k(0)}^{k(t)} F_i^\alpha(\tau)d\tau\right]$$

It now follows in a similar manner as in (c) that $k(\cdot) \in L_\infty(0, \omega)$.

So far we have shown boundedness of $k(\cdot)$ for each feedback strategy $(\alpha) - (\gamma)$ and the proof of the theorem will be completed in the following step (e).

(e): We shall prove that $\omega = \infty$. If (α) is applied, boundedness of $k(\cdot)$ already yields $\omega = \infty$. Consider next the feedback (γ). Although it has been shown that $k(t)$ is bounded, we cannot directly conclude that $\omega = \infty$, since the right hand side of the differential equation (4.10) is discontinuous in k. Suppose $\omega < \infty$. Let $i \in \mathbb{N}$ so that $\omega \in (t_i, t_{i+1}]$ and

$$K_{(Sok)(t)} = K_r \quad \text{for all } t \in [t_i, t_{i+1}) \text{ and some } r \in \underline{N}.$$

Since the right hand side of (4.10) is locally right Lipschitz continuous in (x, k) and since $k(\cdot)$ is bounded, it follows from the classical theory of ordinary differential equations that, for the initial values $k(t_i), x(t_i)$, there exists a unique solution in $[t_i, t_{i+1})$, and hence $\omega = t_{i+1}$. We shall establish that the solution can be extended beyond ω. To this end notice that $\lim_{t \to \omega} k(t) = \tau_j$ for some $j \in \mathbb{N}$. Furthermore, we obtain from the boundedness of $k(\cdot)$ and $\dot{x}(t) = [A - k(t)BK_{(Sok)(t)}C]x(t)$ that $\dot{x}(\cdot) \in L_\infty(0, \omega) \subset L_1(0, \omega)$. Using the fact that

$$x(t) = x(0) + \int_0^t \dot{x}(s)ds \quad \text{for all } t \in [0, \omega)$$

it follows that $\bar{x} := \lim_{t \to \omega} x(t)$ exists. Now the initial value problem

$$\dot{x}(t) = [A + k(t)BK_{S(k(t))}C]x(t) \quad , \quad x(\omega) = \bar{x}$$

$$\dot{k}(t) = \|y(t)\|^p \qquad\qquad , \quad k(\omega) = \tau_j$$

has a unique solution $(x(\cdot), k(\cdot))$ on $[\omega, \omega + \varepsilon)$ for some $\varepsilon > 0$. This extends the solution and therefore, $\omega = \infty$.
In the case of the feedback (β), the proof is analogous, it is omitted. So far we have proved the assertions (i), (ii), and (iv). (iii) is a consequence of Lemma 2.1.8.
This completes the proof. □

In Theorem 4.2.1 we have assumed that (A, B, C) is of relative degree 1. As it was shown in Theorem 2.2.6, there exist systems of relative degree 2 which are high-gain stabilizable. In the following, we are able to present a simple universal adaptive stabilizer for the complete class of positive or negative high-gain stabilizable, single-input, single-output systems with no explicit assumptions on the relative degree. More precisely, we consider the class

$$\left.\begin{array}{l} g(\cdot) \in \mathrm{I\!R}(s), \\[1mm] \text{there exist } k^* > 0 \text{ and } \sigma \in \{-1, +1\} \text{ so that} \\[3mm] \dfrac{g(s)}{1 + k\sigma g(s)} \text{ is exponentially stable for all } k \geq k^* . \end{array}\right\} \quad (4.11)$$

If $g(\cdot)$ is of relative degree 1, a universal adaptive stabilizer is given in Theorem 4.2.1 and the proof is essentially based on the Nussbaum properties (4.3). However, in the relative degree 2 case, similar arguments are not applicable. Instead, we prove directly that an unbounded gain $k(\cdot)$ yields a contradiction as follows. The Nussbaum function $N(k) = \ln k \cos \sqrt{\ln k}$ satisfies the assumptions (2.18) (2.19) for a sequence of intervals $I_1, I_2, \ldots$ and on each interval the state is, by virtue of the High-Gain Lemma 2.2.9, exponentially decaying. Loosely speaking, the intervals become larger and larger. The difference between $N(k) = \ln k \cos \sqrt{\ln k}$ and the standard Nussbaum function $k \cos \sqrt{k}$ is that the logarithm makes the intervals $I_1, I_2, \ldots$ larger. This takes into account the dynamics of an exponentially stable relative degree 2 system, where the oscillation of the state is more rapid than the decrease of its magnitude. Eventually, it is possible to estimate crudely (ignoring that the state was decaying on certain intervals) the whole past of the state, and to show that there exists an interval where $x(t)$ decays to 0 fast enough so that the integral of $y(t)^2$ converges.

Theorem 4.2.2
If

$$\left.\begin{array}{rcl} \dot{x}(t) &=& Ax(t) + bu(t) \qquad x(0) = x_0 \\ y(t) &=& cx(t), \end{array}\right\} \quad (4.12)$$

$(A, b, c) \in \mathbb{R}^{n \times n} \times \mathbb{R}^{n} \times \mathbb{R}^{1 \times n}$, is an arbitrary realization of a transfer function belonging to (4.11), then the feedback strategy

$$
\left.
\begin{aligned}
u(t) &= -\ln k(t) \cos \sqrt{\ln k(t)}\, y(t) \\
\dot{k}(t) &= y(t)^2 \qquad\qquad , \qquad k(0) = k_0
\end{aligned}
\right\} \qquad (4.13)
$$

applied to (4.12), for arbitrary $k_0 > 1$, $x_0 \in \mathbb{R}^n$, yields a closed-loop system with the properties

(i) the unique solution $(x(\cdot), k(\cdot)) : [0, \infty) \to \mathbb{R}^{n+1}$ exists,

(ii) $\lim_{t \to \infty} k(t) = k_\infty$ exists and is finite,

(iii) $x(\cdot) \in L_2(0, \infty) \cap L_\infty(0, \infty)$ and $\lim_{t \to \infty} x(t) = 0$.

Proof: Without loss of generality, we may assume that (4.12) is of one of the special forms (2.3), (2.4). We use the notation $N(k) := \ln k \cos \sqrt{\ln k}$.
(a):   Since the right hand side of the closed-loop system (4.12), (4.13) is continuous and locally Lipschitz in (x, k), existence and uniqueness of a solution $(x(\cdot), k(\cdot)) : [0, \omega) \to \mathbb{R}^{n+1}$ on a maximal interval $[0, \omega)$, $\omega > 0$, follows from the classical theory of differential equations.

(b):   If $cb \neq 0$, then the result follows from Theorem 4.2.1 together with Lemma 4.1.3.

(c):   Suppose $k(\cdot) \in L_\infty(0, \omega)$ and (4.12) is of relative degree 2. It follows from the classical theory of differential equations that $\omega = \infty$. Now the theorem would follow from Lemma 2.1.8.

(d):   It remains to consider the case that (A, b, c) is of relative degree 2 and $k(\cdot) \notin L_\infty(0, \omega)$. Let $M > 0$ be sufficiently large, depending on (A, b, c), so that the assumptions (2.17), (2.18) hold true. Since $N(k(\cdot))$ is unbounded, it takes arbitrarily large negative and positive values and the derivative at the zeros tends to infinity. Therefore, there exists a $p \in \mathbb{N}$ and a sequence

$$
t_{2p} < \hat{t}_{2p} < t_{2p+1} < t_{2(p+1)} < \hat{t}_{2(p+1)} < t_{2(p+1)+1} < \ldots < \omega
$$

such that for all $j \geq p$ we have

$$
\left.
\begin{aligned}
\ln k(t_{2j}) \cos \sqrt{\ln k(t_{2j})} = M \qquad , \qquad \sqrt{\ln k(t_{2j})} \geq (2j - \tfrac{1}{2})\pi \\
\sqrt{\ln k(\hat{t}_{2j})} = (2j - \tfrac{1}{4})\pi \\
\sqrt{\ln k(t_{2j+1})} = 2j\pi.
\end{aligned}
\right\} \qquad (4.14)
$$

By definition of $k(\cdot)$ and and the construction of the intervals $[t_{2j}, t_{2j+1})$, we can apply Lemma 2.2.9. Hence there exist $\hat{M}, \hat{\lambda} > 0$, such that for all $j \geq p$

$$
k(t_{2j+1}) - k(t_{2j}) = \int_{t_{2j}}^{t_{2j+1}} y(s)^2 ds \leq \frac{\hat{M}}{\hat{\lambda}} \|x(t_{2j})\|^2 . \qquad (4.15)
$$

By definition of the systems class (4.11), there exists an $N^* \in \mathbb{R}$ such that

$$\sigma(A^*) \subset \mathbb{C}_- \quad \text{where} \quad A^* := A - N^* bc.$$

The first equation in (4.12) can be written

$$\dot{x}(t) = A^* x(t) - [N(k(t)) - N^*] by(t), \qquad x(0) = x_0$$

and since A^* is exponentially stable there exist $L, \lambda > 0$ such that

$$\|x(t)\| \le L e^{-\lambda t} \|x(0)\| + L\|b\| \left[|N(k(\cdot))|_{L_\infty(0,t)} + |N^*| \right] \int_0^t e^{-\lambda(t-s)} |y(s)| ds.$$
(4.16)

Using Hölder's inequality, we obtain

$$\int_0^t e^{-\lambda(t-s)} |y(s)| ds \;\le\; \left(\int_0^t e^{-\lambda(t-s)} ds \right)^{\frac{1}{2}} \left(\int_0^t e^{-\lambda(t-s)} y(s)^2 ds \right)^{\frac{1}{2}}$$

$$\le \;\; \frac{1}{\sqrt{\lambda}} \left(\int_0^t y(s)^2 ds \right)^{\frac{1}{2}}$$
(4.17)

Taking square roots in (4.15), and inserting (4.16) and (4.17) into (4.15) with

$$M_1 := \sqrt{\frac{\hat{M}}{\omega}} L \|x(0)\|, \qquad M_2 := \sqrt{\frac{\hat{M}}{\omega}} \frac{L\|b\|}{\sqrt{\lambda}}$$

yields

$$\sqrt{k(t_{2j+1}) - k(t_{2j})} \le M_1 + M_2 \left[\|N(k(\cdot))\|_{L_\infty(0,t_{2j})} + |N^*| \right] \sqrt{k(t_{2j}) - k(0)}$$

or equivalently

$$1 \le \frac{M_1}{\sqrt{k(t_{2j+1}) - k(t_{2j})}} + M_2 \left[\|N(k(\cdot))\|_{L_\infty(0,t_{2j})} + |N^*| \right] \sqrt{\frac{k(t_{2j})}{k(t_{2j+1}) - k(t_{2j})}}.$$
(4.18)

In order to derive a contradiction, we show that the right hand side of (4.18) tends to 0 as $j \to \infty$. The first term on the right hand side of (4.18) tends to 0. We shall prove that the second term tends to 0 as well. Using the equations in (4.14) we have

$$\left[\|N(k(\cdot))\|_{L_\infty(0,t_{2j})} + |N^*| \right]^2 \frac{k(t_{2j})}{k(t_{2j+1}) - k(t_{2j})}$$

$$< \;\; \left[4j^2 \pi^2 + |N^*| \right]^2 \frac{k(\hat{t}_{2j})}{e^{4j^2 \pi^2} - k(\hat{t}_{2j})}$$

$$= \;\; \left[4j^2 \pi^2 + |N^*| \right]^2 \frac{e^{(-j+\frac{1}{16})\pi^2}}{1 - e^{(-j+\frac{1}{16})\pi^2}}.$$

The right hand side tends to 0 as $j \to \infty$, thus contradicting (4.18). Therefore, $k(\cdot) \in L_\infty(0,\omega)$. This completes the proof of the theorem.   □

Remark 4.2.3

(i) If for the class of positive or negative high-gain stabilizable systems the sign of the high-frequency gain is known to be positive, then

$$u(t) = -k(t)y(t), \; \dot{k}(t) = y(t)^2, \qquad k(0) \in \mathbb{R}$$

is a universal adaptive controller for the class (4.15). This follows from the proof of Theorem 4.2.2: Part (a)-(c) go through without changes, (d) becomes simpler: If $\lim_{i \to \infty} t_{2j} = \infty$, then there exists a t_{2j} so that the assumptions of Lemma 2.2.9 are satisfied, and hence (4.18) hold true for arbitrary $t_{2j+1} > t_{2j}$. Fixing t_{2j} and choosing $k(t_{2j+1})$ large enough yields a contradiction.

(ii) In Theorem 4.2.2, the derivative of the Nussbaum function tends to 0 as $k \to \infty$, as opposed to diverging derivative for the standard Nussbaum function $N(k) = k \cos \sqrt{k}$. The different switching mechanism takes into account the delayed response of a relative degree 2 system, whose frequency of oscillation is faster than the exponential decay. This feature is emphasized if one changes variables in the gain parameter by $h := \ln(k)$, then

$$u(t) = -h(t) \cos \sqrt{h(t)}\, y(t), \qquad \dot{h}(t) = e^{-h(t)}y(t)^2.$$

We believe, that the feedback strategy $u(t) = k(t) \cos \sqrt{k(t)}\, y(t)$, $\dot{k}(t) = y(t)^2$ is *not* a universal adaptive stabilizer for the class (4.11).

One extension of Theorem 4.2.1 was to consider relative degree 2 systems as we did in Theorem 4.2.2. We can also achieve results for relative degree 0 systems. To this end, we consider almost strictly positive real systems and relate them to minimum phase systems. Theorem 4.2.1 will, in part, be reproved, but the proof will be used in Section 6.1 for robustness results.

Although the simplest adaptive stabilizer for the class of almost strictly positive real systems is $u(t) = -k(t)y(t)$, $\dot{k}(t) = \|y(t)\|^2$, we introduce more complicated controllers since these results will be applied later to minimum phase systems by combining results of Section 3.2 with Theorem 4.2.4.

Theorem 4.2.4
Suppose $G(\cdot) \in GL_m(\mathbb{R}(s))$ is almost strictly positive real with minimal realization

$$\begin{aligned} \dot{x}(t) &= Ax(t) + Bu(t) & x(0) = x_0 \\ y(t) &= Cx(t) + Du(t), \end{aligned} \Bigg\} \tag{4.19}$$

$(A, B, C, D) \in \mathbb{R}^{n \times n} \times \mathbb{R}^{n \times m} \times \mathbb{R}^{m \times n} \times \mathbb{R}^{m \times m}$, and $\mathrm{rk}\, B = m \leq n$. Let

$N(\cdot) : \mathbb{R} \to \mathbb{R}$ be a Nussbaum function, scaling-invariant if $m > 1$,
$\{K_1, \ldots, K_N\}$ a spectrum unmixing set for $GL_m(\mathbb{R})$,
$S(\cdot) : \mathbb{R} \to \underline{N}$ a switching function associated with $\{\tau_i\}_{i \in \mathbb{N}}$ satisfying (4.5),
$\Gamma \in \mathbb{R}^{m \times m}$ so that $\Gamma + \Gamma^T > 0$,
$\Lambda \in \mathbb{R}^{m \times m}$ so that there exists K_i, $i \in \underline{N}$, with $\Lambda K_i + (\Lambda K_i)^T > 0$.

If the adaptation law

$$\dot{k}(t) = \|y(t)\|^2, \qquad k(0) = k_0$$

together with one of the feedback laws

$(\alpha):$ $\quad u(t) = -k(t)\,\Gamma\,y(t)$ $\qquad$, if $\ D + D^T \geq 0, \quad k_0 \geq 0$

$(\beta):$ $\quad u(t) = -N(k(t))\,\Gamma\,y(t)$ $\qquad$, if $\ D = 0$

$(\gamma):$ $\quad u(t) = -k(t)\,\Lambda\,K_{(Sok)(t)}y(t)$ $\quad$, if $\ D = 0$

and arbitrary $k_0 \in \mathbb{R}$, $x_0 \in \mathbb{R}^n$, is applied to (4.19), then the closed-loop system has the properties

(i) the unique solution $(x(\cdot), k(\cdot)) : [0, \infty) \to \mathbb{R}^{n+1}$ exists,

(ii) $\lim_{t \to \infty} k(t) = k_\infty$ exists and is finite,

(iii) $x(\cdot) \in L_2(0, \infty) \cap L_\infty(0, \infty)$ and $\lim_{t \to \infty} x(t) = 0$.

(iv) If (γ) is applied, then there exist $i \in \underline{N}$ and $t^* > 0$, so that $K_{S(k(t))} = K_i$ for all $t \geq t^*$.

Proof: (a): Since $\Gamma + \Gamma^T > 0$ it follows from Remark 3.1.4 (iv) that $\Gamma^{-1} + \Gamma^{-T} > 0$, and hence, in view of $D + D^T \geq 0$,

$$(\Gamma^{-1} + kD) + (\Gamma^{-1} + kD)^T > 0 \quad \text{for all} \quad k \geq 0,$$

which yields

$$|I_m + kD\Gamma| = |\Gamma^{-1} + kD| \cdot |\Gamma| \neq 0 \quad \text{for all} \quad k \geq 0.$$

Therefore, the initial value problem for the feedback (α)

$$\dot{x}(t) \;=\; [A - k(t)B\Gamma[I_m + k(t)D\Gamma]^{-1}C]\,x(t) \quad, \quad x(0) = x_0$$

$$\dot{k}(t) \;=\; \|\,[I_m + k(t)D\Gamma]^{-1}Cx(t)\|^2 \qquad\qquad , \quad k(0) = k_0$$

is well defined. It now follows as in part (a) of the proof of Theorem 4.2.1 that, in all cases (α)-(γ), there exists has a unique solution $(x(\cdot), k(\cdot)) : [0, \omega) \to \mathbb{R}^{n+1}$, maximally extended over $[0, \omega)$ for some $\omega > 0$.
By assumption, there exists a $K \in \mathbb{R}^{m \times m}$ such that the closed-loop system $H(s) = [I_m + G(s)K]^{-1}G(s)$ is strictly positive real.

(b):   Consider the feedback $u(t) = -k(t)\Gamma y(t)$. Applying $u(t) = -Ky(t) - (k(t)\Gamma - K)y(t)$ in Lemma 3.1.6 yields, for all $s \in [0, \omega)$,

$$\frac{d}{ds}V(x(s)) \leq -2\mu V(x(s)) - \gamma k(t)\|y(s)\|^2 + 2\|K\|\|y(s)\|^2$$

for some $\mu > 0$ and $\gamma := \mu_{\min}(\Gamma + \Gamma^T)$. Integration over $[0, t]$ and changing variables, $\tau = k(s)$, gives

$$\begin{aligned}
V(x(t)) &\leq V(x(0)) - \int_0^t [\gamma k(s) - 2\|K\|]\|y(s)\|^2 ds \\
&= V(x(0)) - [k(t) - k(0)] \cdot \left[\frac{\gamma}{2}[k(t) + k(0)] - 2\|K\|\right]
\end{aligned}$$

for all $t \in [0, \omega)$. If $k(\cdot) \notin L_\infty(0, \omega)$, then the right hand side gets negative, hence contradicting the non-negativeness of the left hand side. This proves $k(\cdot) \in L_\infty(0, \omega)$.

(c):   Consider the feedback

$$u(t) = -Ky(t) - (\mathcal{K}(t) - K)y(t), \qquad \mathcal{K}(t) = k(t)\Lambda K_{(Sok)(t)}.$$

Lemma 3.1.6 yields, for some $K \in \mathbb{R}^{m \times m}$, $\mu > 0$, and all $t \in [0, \omega)$,

$$\begin{aligned}
\frac{d}{dt}V(x(t)) &\leq -2\mu V(x(t)) - k(t)\left\langle y(t), \left[\Lambda K_{(Sok)(t)} + (\Lambda K_{(Sok)(t)})^T\right] y(t)\right\rangle \\
&\quad +2\|K\|\|y(t)\|^2.
\end{aligned} \tag{4.20}$$

By assumption, there exists a $K_i \in \{K_1, \ldots, K_N\}$ so that

$$\Lambda K_i + (\Lambda K_i)^T > 0,$$

and hence

$$-\left\langle y(t), \left[\Lambda K_i + (\Lambda K_i)^T\right] y(t)\right\rangle \leq -\gamma\|y(t)\|^2$$

for $\gamma := \mu_{\min}\left(\Lambda K_i + (\Lambda K_i)^T\right) > 0$. Let $\alpha > 0$ so that

$$\Lambda K_j + (\Lambda K_j)^T \geq -\gamma\alpha I_m \qquad \text{for all} \quad j \in \underline{N}.$$

Defining F_i^α as in (4.6), it follows that

$$-k(t)\left\langle y(t), \left[\Lambda K_{(Sok)(t)} + (\Lambda K_{(Sok)(t)})^T\right] y(t)\right\rangle \leq -\gamma F_i^\alpha(k(t))\|y(t)\|^2$$

and hence, by integrating $\frac{d}{ds}V(x(s))$ over $[0, t]$ and changing variables, we obtain

$$\begin{aligned}
V(x(t)) &\leq V(x(0)) - \int_0^t \gamma F_i^\alpha(k(s))\|y(s)\|^2 ds + 2\|K\| \int_0^t \|y(s)\|^2 ds \\
&= V(x(0)) - \int_{k(0)}^{k(t)} \gamma F_i^\alpha(\tau)d\tau + 2\|K\|[k(t) - k(0)]
\end{aligned}$$

for all $t \in [0, \omega)$. This gives

$$V(x(t)) \leq V(x(0)) + [k(t) - k(0)] \left[2\|K\| - \frac{\gamma}{k(t) - k(0)} \int_{k(0)}^{k(t)} F_i^\alpha(\tau) d\tau \right]$$

for all $t \in [0, \omega)$ so that $k(t) > k(0)$. If $k(\cdot) \notin L_\infty(0, \omega)$, then it follows from Lemma 4.1.7 that the right hand side of the above inequality takes negative values, thus contradiction the non-negativeness of $V(x(t))$. Therefore, $k(\cdot) \in L_\infty(0, \omega)$.

(d): It is simpler to derive the same conclusion in the case $u(t) = -N(k(t))y(t)$. Instead of (4.20) we obtain

$$\frac{d}{dt} V(x(t)) \leq -N(k(t)) \langle y(t), [\Gamma + \Gamma^T] y(t) \rangle + 2\|K\| \|y(t)\|^2$$

$$\leq \left(\tilde{N}(k(t)) + 2\|K\| \right) \|y(t)\|^2$$

where

$$\tilde{N}(k) := \begin{cases} -2\|\Gamma\| N(k) & , \text{ if } N(k) < 0 \\ -\mu_{\min} \left(\Gamma + \Gamma^T \right) N(k) & , \text{ if } N(k) \geq 0. \end{cases}$$

Integrating $\frac{d}{ds} V(x(s))$ over $[0, t]$ and changing variables again yields

$$V(x(t)) \leq V(x(0)) + \int_{k(0)}^{k(t)} \tilde{N}(\tau) d\tau + 2\|K\| [k(t) - k(0)]$$

$$= V(x(0)) + [k(t) - k(0)] \left[2\|K\| + \frac{1}{k(t) - k(0)} \int_{k(0)}^{k(t)} \tilde{N}(\tau) d\tau \right]$$

for all $t \in [0, \omega)$ so that $k(t) > k(0)$. Since $N(k)$ is assumed to be scaling-invariant, $\tilde{N}(k)$ is a Nussbaum function as well, and it follows that the right hand side of the above inequality takes negative values if $\lim_{t \to \omega} k(t) = +\infty$, thus contradiction the non-negativeness of $V(x(t))$. This proves again $k(\cdot) \in L_\infty(0, \omega)$.

(e): Since $k(\cdot) \in L_\infty(0, \omega)$, it follows from the inequalities of $V(x(t))$ in (b)-(d) that $V(x(\cdot)) \in L_\infty(0, \omega)$, and hence $x(\cdot) \in L_\infty(0, \omega)$. This proves the assertions (i), (ii), and (iv). For $u(t)$ written as $u(t) = -\mathcal{K}(t)y(t)$, depending on which feedback is chosen, $x(t)$ satisfies, confer (3.13),

$$\dot{x}(t) = \hat{A}x(t) - \hat{B} \left(\mathcal{K}(t) - K \right) y(t).$$

Since $\lim_{t \to \infty} \mathcal{K}(t)$ is finite and $\hat{A}$ is asymptotically stable, $y(\cdot) \in L_2(0, \infty)$ yields $x(\cdot) \in L_2(0, \infty)$, and hence $\dot{x}(\cdot) \in L_2(0, \infty)$. Now (iii) is a consequence of Lemma 2.1.7. This completes the proof. □

The previous result is now applied to obtain a universal adaptive stabilizer for strictly proper minimum phase systems. A result for proper, but not strictly proper, systems is given in Theorem 4.2.6.

Corollary 4.2.5
Suppose $G(\cdot) \in GL_m(\mathbb{R}(s))$ is strictly proper and minimum phase with minimal realization

$$\left.\begin{array}{rcl} \dot{x}(t) & = & Ax(t) + Bu(t) \qquad , x(0) = x_0 \\ y(t) & = & Cx(t), \end{array}\right\} \tag{4.21}$$

$(A, B, C) \in \mathbb{R}^{n \times n} \times \mathbb{R}^{n \times m} \times \mathbb{R}^{m \times n}$, and $\mathrm{rk} B = m \leq n$. Let

$N(\cdot) : \mathbb{R} \to \mathbb{R}$ be a Nussbaum function, scaling-invariant if $m > 1$,

$\{K_1, \ldots, K_N\}$ a spectrum unmixing set for $GL_m(\mathbb{R})$,

$S(\cdot) : \mathbb{R} \to \underline{N}$ a switching function associated with $\{\tau_i\}_{i \in \mathbb{N}}$ satisfying (4.5).

If the adaptation law

$$\dot{k}(t) = \|y(t)\|^2, \qquad k(0) = k_0$$

together with one of the feedback laws

$(\alpha):\qquad u(t) = -k(t)y(t) \qquad\qquad , \text{ if } \sigma(CB) \subset \mathbb{C}_+$

$(\beta):\qquad u(t) = -N(k(t))y(t) \qquad , \text{ if } \sigma(CB) \subset \mathbb{C}_+ \text{ or } \mathbb{C}_-$

$(\gamma):\qquad u(t) = -k(t)K_{(S \circ k)(t)}y(t) \quad , \text{ if } |CB| \neq 0$

and arbitrary $k_0 \in \mathbb{R}$, $x_0 \in \mathbb{R}^n$, is applied to (4.21), then the closed-loop system has the properties

(i) the unique solution $(x(\cdot), k(\cdot)) : [0, \infty) \to \mathbb{R}^{n+1}$ exists,

(ii) $\lim_{t \to \infty} k(t) = k_\infty$ exists and is finite,

(iii) $x(\cdot) \in L_2(0, \infty) \cap L_\infty(0, \infty)$ and $\lim_{t \to \infty} x(t) = 0$.

(iv) If (γ) is applied, then there exist $i \in \underline{N}$ and $t^* > 0$, so that $K_{S(k(t))} = K_i$ for all $t \geq t^*$.

Proof: (a):   First we prove the theorem for the feedback (β). Let $R = R^T \in \mathbb{R}^{m \times m}$ be the positive-definite solution of

$$RCB + (CB)^T R = \sigma I_m$$

for some $\sigma \in \{-1, +1\}$. If $U := RCB$, then, by Proposition 3.2.1, the system

$$\begin{array}{rcl} \dot{x}(t) & = & Ax(t) + BU^{-1}\bar{u}(t) \\ y(t) & = & Cx(t) \end{array}$$

is almost strictly positive real and

$$\bar{u}(t) = Uu(t) = -RCBN(k(t))y(t) = -\sigma N(k(t))\Gamma y(t)$$

with $\Gamma := \sigma RCB$. Since Γ satisfies the assumptions of Theorem 4.2.4, the assertions (i)-(iii) follow.

(b): The proof for the feedback (α) follows by setting $N(k) = k$ and $\sigma = 1$ in part (a).

(c): It remains to prove the theorem for the feedback (γ). Choose $K_i \in \{K_1, \ldots, K_N\}$ so that $\sigma(CBK_i) \subset \mathbb{C}_+$. Hence there exists a positive-definite $R = R^T \in \mathbb{R}^{m \times m}$ solving

$$RCBK_i + (CBK_i)^T R = I_m.$$

If $U := RCB$, then by Proposition 3.2.1

$$\begin{aligned}
\dot{x}(t) &= Ax(t) + BU^{-1}\bar{u}(t) \\
y(t) &= Cx(t)
\end{aligned}$$

is almost strictly positive real and

$$\bar{u}(t) = Uu(t) = -k(t)RCBK_{(Sok)(t)}y(t).$$

For $\Lambda = RCB$ the result follows from Theorem 4.2.4.
This completes the proof. $\qquad\square$

It is known that there exist systems (A, B, C, D) such that D is singular but not zero and which are high-gain stabilizable via $u(t) = -ky(t)$ with constant k. In the following theorem, we introduce a class of systems for which this result can be achieved adaptively.

Theorem 4.2.6
Suppose $G(\cdot) \in GL_m(\mathbb{R}(s))$ is minimum phase with minimal realization

$$\begin{aligned}
\dot{x}(t) &= Ax(t) + Bu(t) \quad , x(0) = x_0 \\
y(t) &= Cx(t) + Du(t),
\end{aligned} \right\} \qquad (4.22)$$

where $(A, B, C, D) \in \mathbb{R}^{n \times n} \times \mathbb{R}^{n \times m} \times \mathbb{R}^{m \times n} \times \mathbb{R}^{m \times m}$, and $\operatorname{rk} B = m \le n$.
Suppose furthermore, (4.22) satisfies (I)-(III) below:

(I) There exists an orthogonal matrix $[Z, W]$, $Z \in \mathbb{R}^{m \times (m-r)}$, $W \in \mathbb{R}^{m \times r}$, $r = \operatorname{rk} D$, such that

$$[Z, W]^T D[Z, W] = \begin{bmatrix} 0, & 0 \\ 0, & W^T DW \end{bmatrix} \qquad (4.23)$$

(II) $\det(Z^T C B Z) \neq 0$,

(III) $W^T[D + D^T]W \geq 0$.

Let

$N(\cdot) : \mathbb{R} \to \mathbb{R}$ be a Nussbaum function, scaling-invariant if $m - r > 1$,

$\{K_1, \ldots, K_N\}$ a spectrum unmixing set for $GL_{m-r}(\mathbb{R})$,

$S(\cdot) : \mathbb{R} \to \underline{N}$ a switching function associated with $\{\tau_i\}_{i \in \mathbb{N}}$ satisfying (4.5).

If the adaptation law

$$\dot{k}(t) = \|y(t)\|^2, \qquad k(0) = k_0$$

respectively

$$\dot{k}_1(t) = \|Z^T y(t)\|^2, \quad k_1(0) = k_0^1, \quad \text{and} \quad \dot{k}_2(t) = \|W^T y(t)\|^2, \quad k_2(0) = k_0^2,$$

together with one of the feedback laws

$(\alpha): \quad u(t) = -k(t)y(t)$ if $\sigma(Z^T C B Z) \subset \mathbb{C}_+$

$(\beta): \quad u(t) = -[Z, W]\begin{bmatrix} N(k_1(t))I_{m-r}, & 0 \\ 0 & k_2(t)I_r \end{bmatrix}[Z, W]^T y(t)$

if $\sigma(Z^T C B Z) \subset \mathbb{C}_+$ or $\mathbb{C}_-$

$(\gamma): \quad u(t) = -[Z, W]\begin{bmatrix} k_1(t)K_{(S \circ k_1)(t)}, & 0 \\ 0 & k_2(t)I_r \end{bmatrix}[Z, W]^T y(t)$

if $|Z^T C B Z| \neq 0$

and arbitrary $k_0, k_0^2 \geq 0$, $k_0^1 \in \mathbb{R}$, $x_0 \in \mathbb{R}^n$, is applied to (4.22), then the closed-loop system has the properties

(i) the unique solution $(x(\cdot), k(\cdot)) : [0, \infty) \to \mathbb{R}^{n+1}$, resp. $(x(\cdot), k_1(\cdot), k_2(\cdot)) : [0, \infty) \to \mathbb{R}^{n+2}$, exists,

(ii) $\lim_{t \to \infty} k(t) = k_\infty$ resp. $\lim_{t \to \infty} k_i(t) = k_\infty^i$, $i = 1, 2$, exist and are finite,

(iii) $x(\cdot) \in L_2(0, \infty) \cap L_\infty(0, \infty)$ and $\lim_{t \to \infty} x(t) = 0$.

(iv) If (γ) is applied, then there exist $i \in \underline{N}$ and $t^* > 0$, so that $K_{S(k_1(t))} = K_i$ for all $t \geq t^*$.

Proof: (a):   First, it is established that the feedback system is well defined. To this end define

$$\bar{y}(t) = [Z, W]^T y(t), \qquad \bar{y}_1(t) = Z^T y(t), \qquad \bar{y}_2(t) = W^T y(t)$$

and let

$$\mathcal{K}_1(t) = k_1(t)I_{m-r}, \quad = N(k_1(t))I_{m-r}, \quad \text{or} \quad = k_1(t)K_{S(k_1(t))}.$$

It is easy to see that the second equation in (4.22) is equivalent to

$$\begin{aligned}
\bar{y}_1(t) &= Z^T Cx(t) \\
\bar{y}_2(t) &= W^T Cx(t) - W^T DW k_2(t)\bar{y}_2(t).
\end{aligned}$$

Since $W^T[D + D^T]W \geq 0$ and $k_2(t) \geq 0$, it follows (confer the proof of Proposition 3.1.4) that

$$\bar{y}_2(t) = \left[I_r + k_2(t)W^T DW\right]^{-1} W^T Cx(t).$$

Inserting $\bar{y}_1(t)$ and $\bar{y}_2(t)$ into the first equation in (4.22) yields

$$\dot{x}(t) = \left\{A - B[Z, W] \begin{bmatrix} \mathcal{K}_1(t)Z^T \\ (I_r + k_2(t)W^T DW)^{-1}W^T \end{bmatrix} C \right\} x(t).$$

As in part (a) of the proof of Theorem 4.2.1, it follows that, in case of (α), for every $(x_0, k_0) \in \mathbb{R}^{n+1}$, the closed-loop system has a unique solution $(x(\cdot), k(\cdot))$: $[0, \omega) \to \mathbb{R}^{n+1}$, maximally extended over $[0, \omega)$ for some $\omega > 0$. The same is valid for $(x(\cdot), k_1(\cdot), k_2(\cdot)) : [0, \omega) \to \mathbb{R}^{n+2}$ if the switching strategy (β) or (γ) is chosen.

(b):   Consider the feedback (β). Let $R = R^T \in \mathbb{R}^{(m-r)\times(m-r)}$ be the positive-definite solution of

$$R Z^T CBZ + (Z^T CBZ)^T R = \sigma I_{m-r}.$$

for some $\sigma \in \{-1, +1\}$. By Theorem 3.2.3, the system (4.22) is equivalent to the almost strictly positive real system (3.22) with

$$\bar{u}(t) = S[Z, W]^T u(t) = -S \begin{bmatrix} N(k_1(t))I_{m-r}, & 0 \\ 0 & k_2(t)I_r \end{bmatrix} \bar{y}(t),$$

where $S = \text{diag}\{R Z^T CBZ, I_r\}$. It follows from Lemma 3.1.6, applied to (3.22), that

$$\frac{d}{dt} V(x(t)) \leq -2 \left\langle \bar{y}(t), \begin{bmatrix} \sigma N(k_1(t))I_{m-r}, & 0 \\ 0 & k_2(t)I_r \end{bmatrix} \bar{y}(t) \right\rangle + 2\|K\|\|\bar{y}(t)\|^2$$

for some $K \in \mathbb{R}^{m\times m}$ and positive-definite $V(x) = \langle x, Px \rangle$ as defined in Lemma 3.1.6, and all $t \in [0, \omega)$. Integrating $\frac{d}{ds} V(x(s))$ over $[0, t]$ yields

$$V(x(t)) \leq V(x(0)) - 2 \int_0^t [\sigma N(k_1(s)) - \|K\|] \|\bar{y}_1(s)\|^2 ds$$

$$-2 \int_0^t [k_2(s)] - \|K\|] \|\bar{y}_2(s)\|^2 ds$$

$$= V(x(0)) - 2 \int_{k_1(0)}^{k_1(t)} [\sigma N(\tau) - \|K\|] d\tau - 2 \int_{k_2(0)}^{k_2(t)} [\tau - \|K\|] d\tau.$$

If $k_1(\cdot) \notin L_\infty(0,\omega)$ or $k_2(\cdot) \notin L_\infty(0,\omega)$, then the right hand side of the above inequality takes negative values, thus contradicting the non-negativeness of the left hand side. Therefore, $k_1(\cdot), k_2(\cdot) \in L_\infty(0,\omega)$ and hence $y(\cdot) \in L_2(0,\omega)$. This yields $\omega = \infty$, and the assertions (i), (ii) are proved for the feedback (β). Since $\hat{A}$ in (3.15) is exponentially stable, it follows that $x(\cdot) \in L_2(0,\infty)$, whence $\dot{x}(\cdot) \in L_2(0,\infty)$, and hence (iii) follows from Lemma 4.1.1.

(c):   The proof for (α) is analogous to (b).

(d):   It remains to consider the feedback (γ). We shall establish that $k_1(\cdot),\ k_2(\cdot) \in L_\infty(0,\omega)$. To this end, choose $K_i \in \{K_1,\dots,K_N\}$ such that $\sigma(Z^T CBZK_i) \subset \mathbb{C}_+$ and let $R = R^T \in \mathbb{R}^{(m-r)\times(m-r)}$ be the positive-definite solution of

$$R Z^T CBZK_i + (Z^T CBZK_i)^T R = I_{m-r}.$$

Define $\Lambda = RZ^T CBZ$, and choose $\alpha > 0$ sufficiently large so that

$$-\Lambda K_j - (\Lambda K_j)^T \leq \alpha I_{m-r} \quad \text{for all} \quad j \in \underline{N}.$$

Again, the system (4.22) is equivalent to the almost strictly positive real system (3.22) and, by Lemma 3.1.6, it follows that

$$\frac{d}{dt}V(x(t)) \leq -\left\langle \bar{y}(t), \left[\begin{array}{cc} k_1(t)\left[\Lambda K_{s(k(t))} + (\Lambda K_{s(k(t))})^T\right], & 0 \\ 0 & k_2(t)I_r \end{array} \right] \bar{y}(t) \right\rangle$$

$$+2\|K\| \|\bar{y}(t)\|^2 .$$

for positive-definite $V(x) = \langle x, Px \rangle$ as defined in Lemma 3.1.6. Using $F_\alpha^i(\cdot)$ defined in (4.6), and assuming without loss of generality that $i = N$, we obtain

$$\frac{d}{dt}V(x(t)) \leq -F_\alpha^i(k_1(t))\|\bar{y}_1(t)\|^2 - k_2(t)\|\bar{y}_2(t)\|^2 + 2\|K\|\|y(t)\|^2$$

$$= -[F_\alpha^i(k_1(t)) - 2\|K\|] \|\bar{y}_1(t)\|^2 - [k_2(t) - 2\|K\|] \|\bar{y}_2(t)\|^2.$$

It now follows in a similar manner as in the proof of part (b) that $k_1(\cdot)$, $k_2(\cdot) \in L_\infty(0,\omega)$.

By the same arguments as in part (e) of the proof of Theorem 4.2.1 we obtain
$\omega = \infty$, and hence the assertions (i), (ii), and (iv) are valid for the feedback
(γ). (iii) follows from Lemma 2.1.8.
This completes the proof. $\square$

Remark 4.2.7
We cannot see how to design a universal adaptive stabilizer for a system
(A, B, C, D) where the assumption $D + D^T \geq 0$ is not made and $D \neq 0$.
Consider, for example, the simple scalar, minimum phase system

$$\begin{aligned}
\dot{x}(t) &= -u(t) &, \; x(0) = x_0 \in \mathbb{R} \\
y(t) &= x(t) - u(t).
\end{aligned}$$

If it is assumed that $d = -1$ is unknown and therefore a feedback strategy

$$\begin{aligned}
\dot{k}(t) &= y(t)^2 &, k(0) = 0 \\
u(t) &= -N(k(t))y(t) &= -k(t)^2 \cos k(t)\, y(t)
\end{aligned}$$

is chosen, then

$$\dot{x}(t) = \frac{N(k(t))}{1 - N(k(t))} x(t) \quad , \; x(0) = x_0$$

$$y(t) = \frac{1}{1 - N(k(t))} x(t)$$

is well defined as long as $N(k(t)) < 1$. Therefore, $x(t)$ does not have a finite
escape time, however $y(t)$ has as we shall show.
Suppose $\lim_{t\to\infty} k(t) = k_\infty$ exists and is finite and $\lim_{t\to\infty} x(t) = 0$. Since

$$\frac{1}{2}\frac{d}{dt}x(t)^2 = x(t)\dot{x}(t) = \frac{1}{1 - N(k(t))}x(t)^2 = \frac{[1 - N(k(t))]^2}{1 - N(k(t))}y(t)^2,$$

integration over $[0, t]$ and changing variables yields

$$\frac{1}{2}x(t)^2 - \frac{1}{2}x(0)^2 = \int_{k(0)}^{k(t)} N(\tau)[1 - N(\tau)]d\tau,$$

and thus, for $t \to \infty$, we obtain

$$-\frac{1}{2}x_0^2 = \int_0^{k_\infty} N(\tau)[1 - N(\tau)]d\tau.$$

x_0 large implies that k_∞ is large, and hence for x_0 large, there exist $0 < t_1 < t_2$
such that $N(k(t_1)) = N(k(t_2)) = 1$. This yields $x(t_1) \neq 0$ since otherwise

$x(t) = 0$ for all $t \geq t_1$, and hence $k(t_1) = k(t)$ for all $t \geq t_1$. But then it follows that $\lim_{t \nearrow t_1} y(t) = +\infty$, therefore the closed-loop system is not well defined.

Remark 4.2.8

The control strategy

$$
\begin{aligned}
\dot{k}(t) &= \|y(t)\|^p \\
u(t) &= -N(k(t))y(t)
\end{aligned}
$$

is a special case of

$$
\begin{aligned}
\dot{k}(t) &= \kappa(t)\|y(t)\|^p \\
u(t) &= -N(k(t))\kappa(t)y(t),
\end{aligned}
$$

where $\kappa(\cdot) : \mathbb{R} \to \mathbb{R}_+$ is a continuous map with

$$
\kappa(t) \geq \varepsilon > 0 \quad \text{for some} \quad \varepsilon > 0.
$$

It is easy to see that all previous results of the present section remain valid for this modified switching strategy. Depending on the choice of $\kappa(\cdot)$, which may be the solution of a differential equation, the controller becomes more complex on the benefit of a large flexibility which might be used by the designer for improving the performance.

4.3 Stabilization via switching decision function

In this section, an alternative adaptive switching strategy to the Nussbaum-type strategy introduced in Section 4.2 is given. It is based on the switching decision function introduced in Section 4.1. The advantage of this different approach is that we do not have to incorporate a *scaling-invariant* Nussbaum function (which grows rapidly), since the times when switching occurs is more directly determined by the system dynamics.

Theorem 4.3.1

Suppose the system

$$
\left.
\begin{aligned}
\dot{x}(t) &= Ax(t) + Bu(t) \quad , \quad x(0) = x_0 \\
y(t) &= Cx(t),
\end{aligned}
\right\}
\tag{4.24}
$$

with $(A, B, C) \in \mathbb{R}^{n \times n} \times \mathbb{R}^{n \times m} \times \mathbb{R}^{m \times n}$, is minimum phase and $\sigma(CB) \subset \mathbb{C}_+$ or $\mathbb{C}_-$. Let $0 < \lambda_1 < \lambda_2 < \ldots$ be a strictly increasing sequence with $\lim_{i \to \infty} \lambda_i = \infty$.

If the feedback strategy

$$
\begin{aligned}
\dot{k}(t) &= \|y(t)\|^2 &, k(0) = k_0 \\
u(t) &= -k(t)\Theta(t)y(t),
\end{aligned}
$$

for arbitrary $k_0 \in \mathbb{R}, x_0 \in \mathbb{R}^n$, is applied to (4.24) via

$$
\Theta(\cdot) : [0, \infty) \rightarrow \{-1, +1\}
$$

where $\Theta(\cdot)$ is determined by the switching decision function

$$
\psi(t) = \frac{k_0 + \int\limits_0^t \Theta(\tau)k(\tau)\|y(\tau)\|^2 d\tau}{1 + \int\limits_0^t \|y(\tau)\|^2 d\tau}
$$

and the algorithm

$$
(*) \quad
\left.
\begin{aligned}
i &:= 0 \\
\Theta(0) &:= -1 \quad, \quad t_0 = 0 \\
t_{i+1} &:= \inf\{t > t_i | \, |\psi(t)| \le \lambda_{i+1} k(0)\} \\
\Theta(t) &:= \Theta(t_i) \quad \text{for all} \quad t \in [t_i, t_{i+1}) \\
\Theta(t_{i+1}) &:= -\Theta(t_i) \\
i &:= i+1 \\
&\text{go to } (*)
\end{aligned}
\right\}
\quad (4.25)
$$

then the closed-loop system has the properties

(i) the unique solution $(x(\cdot), k(\cdot)) : [0, \infty) \rightarrow \mathbb{R}^{n+1}$ exists,

(ii) $\lim_{t \to \infty} k(t) = k_\infty$ exists and is finite,

(iii) $x(\cdot) \in L_2(0, \infty) \cap L_\infty(0, \infty)$ and $\lim_{t \to \infty} x(t) = 0$,

(iv) $\lim_{t \to \infty} \psi(t) = \psi_\infty$ exists and is finite and thus only a finite number of switches occur.

Proof: (a): The closed-loop system is given by

$$
\begin{aligned}
\dot{x}(t) &= [A - k(t)\Theta(t)BC]\, x(t) &, \ x(0) = x_0 \\
\dot{k}(t) &= \|Cx(t)\|^2 &, \ k(0) = k_0.
\end{aligned}
$$

Since, by construction, $\Theta(t)$ is piecewise right continuous, the right hand side of the above differential equation is not continuous in x, k, but locally right

Lipschitz continuous in x, k. It follows by the classical theory of ordinary differential equations that for every $(x_0, k_0) \in \mathbb{R}^{n+1}$, the closed-loop system has a unique solution $(x(\cdot), k(\cdot)) : [0, \omega) \to \mathbb{R}^{n+1}$, maximally extended over $[0, \omega)$ for some $\omega > 0$.

(b):   We shall establish that $k(\cdot) \in L_\infty(0, \infty)$. Choose $R = R^T \in \mathbb{R}^{m \times m}$ to be the positive-definite solution of

$$RCB + (CB)^T R = 2\sigma I_m$$

for some $\sigma \in \{-1, +1\}$. The first inequality in Lemma 2.1.6 yields

$$\frac{1}{2}\|y(t)\|^2 \leq M\|x(0)\|^2 + M \int_0^t \|y(s)\|^2 ds + \sigma \int_0^t \Theta(\tau)k(\tau)\|y(\tau)\|^2 d\tau$$

$$= M\|x(0)\|^2 - M + M \left[1 + \int_0^t \|y(s)\|^2 ds\right] - \sigma k_0$$

$$+ \sigma \left[k_0 + \int_0^t \Theta(\tau)k(\tau)\|y(\tau)\|^2 d\tau\right]$$

$$\leq M\|x(0)\|^2 - \sigma k_0 + [1 + k(t) - k(0)] [M + \sigma\psi(t)] .$$

If $k(\cdot) \notin L_\infty(0, \omega)$, then, in view of Lemma 4.1.9, the right hand side of the above inequality takes negative values, which is a contradiction.

(c)   So far we have shown $k(\cdot) \in L_\infty(0, \omega)$ and $\psi(\cdot) \in L_\infty(0, \omega)$. This implies $\omega = \infty$ since otherwise the solution could be extended beyond ω, cf. part (e) of the proof of Theorem 4.2.1. This proves the assertions (i), (ii) and (iv). (iii) follows from Lemma 2.1.8.
This completes the proof of the theorem.   □

Remark 4.3.2
The adaptation law in Theorem 4.3.1 can be generalized to

$$\dot{k}(t) = \|y(t)\|^p, \qquad k(0) = k_0$$

for arbitrary $p \geq 1$, and $\psi(t)$ as defined in Section 4.1. For single-input, single-output systems, the inequality in part (b) of the proof of Theorem 4.3.1 is then replaced by an inequality derived from the second inequality in Lemma 2.1.6. The remainder goes through analogously. For multivariable systems, a different and more complicated proof is given in Ilchmann and Owens (1991a).

It will be seen in Chapter 6, that the above switching strategy can cope with a large class of nonlinear disturbances.

In the following theorem it will be shown that stabilization via a switching decision function is also applicable for a class of proper, but not necessarily strictly proper, minimum phase systems.

Theorem 4.3.3
Suppose $G(\cdot) \in GL_m(\mathbb{R}(s))$ is minimum phase with minimal realization

$$\left. \begin{array}{rll} \dot{x}(t) & = & Ax(t) + Bu(t) \\ y(t) & = & Cx(t) + Du(t) \end{array} \quad , \quad x(0) = x_0 \right\} \qquad (4.26)$$

$(A, B, C, D) \in \mathbb{R}^{n \times n} \times \mathbb{R}^{n \times m} \times \mathbb{R}^{m \times n} \times \mathbb{R}^{m \times m}$, and $\mathrm{rk}\, B = m \le n$.
Suppose furthermore, (4.26) satisfies (I)-(III) below:

(I) There exists an orthogonal matrix $[Z, W]$, $Z \in \mathbb{R}^{m \times (m-r)}$, $W \in \mathbb{R}^{m \times r}$, $r = \mathrm{rk}\, D$ such that

$$[Z, W]^T D[Z, W] = \begin{bmatrix} 0, & 0 \\ 0, & W^T DW \end{bmatrix} \qquad (4.27)$$

(II) $\sigma(Z^T CBZ) \subset \mathbb{C}_+$ or $\mathbb{C}_-$,

(III) $W^T[D + D^T]W \ge 0$.

If the adaptation law

$$\dot{k}_1(t) = \|Z^T y(t)\|^2, \quad k_1(0) = k_0^1, \quad \text{and} \quad \dot{k}_2(t) = \|W^T y(t)\|^2, \quad k_2(0) = k_0^2,$$

together with the feedback law

$$u(t) = -[Z, W] \begin{bmatrix} k_1(t)\Theta_1(t)I_{m-r}, & 0 \\ 0 & k_2(t)I_r \end{bmatrix} [Z, W]^T y(t),$$

where $\Theta_1(\cdot) : [0, \infty) \to \{-1, +1\}$ is determined via

$$\psi_1(t) = \frac{k_1(0) + \int\limits_0^t \Theta_1(\tau)k_1(\tau)\|Z^T y(\tau)\|^2 d\tau}{1 + \int\limits_0^t \|Z^T y(\tau)\|^2 d\tau}$$

and the algorithm (4.25) with $\Theta(t)$ replaced by $\Theta_1(t)$, is applied to (4.26), then the properties (i)-(iv) in Theorem 4.3.1 are valid as well.

Proof: (a):   Well posedness of the closed-loop system and existence and uniqueness of a solution $(x(\cdot), k_1(\cdot), k_2(\cdot)) : [0, \omega) \to \mathbb{R}^{n+2}$, maximally extended over $[0, \omega)$, for some $\omega > 0$, follows as in part (a) of the proofs of Theorem 4.2.6 and 4.3.1.

(b):   In a similar manner as in part (b) of the proof of Theorem 4.2.6 we
conclude, using the same notation introduced therein, for all $t \in [0, \omega)$,

$$\frac{d}{dt}V(x(t)) \leq -2\left[\sigma k_1(t)\Theta_1(t)\|\bar{y}_1(t)\|^2 + k_2(t)\|\bar{y}_2(t)\|^2\right] + 2\|K\|\|\bar{y}(t)\|^2$$

Integration over $[0, t]$ yields

$$
\begin{aligned}
V(x(t)) \quad &\leq \quad V(x(0)) - 2\int_0^t [\sigma k_1(\tau)\Theta_1(\tau) - \|K\|] \|\bar{y}_1(\tau)\|^2 d\tau \\
&\quad -2\int_0^t [k_2(\tau) - \|K\|] \|\bar{y}_2(\tau)\|^2 d\tau \\
&= \quad V(x(0)) + 2[\sigma k_0^1 - \|K\|] + [1 + k_1(t) - k_1(0)][\|K\| - \sigma\psi_1(t)] \\
&\quad -[k_2(t) - k_2(0)][k_2(t) + k_2(0) - 2\|K\|] .
\end{aligned}
$$

If $k_1(\cdot) \notin L_\infty(0, \omega)$ or $k_2(\cdot) \notin L_\infty(0, \omega)$, then the right hand side of the above
inequality takes negative values, thus contradicting the non-negativeness of the
left hand side. Therefore, $k_1(\cdot), k_2(\cdot) \in L_\infty(0, \omega)$.

(c):   Boundedness of $k_1(\cdot), k_2(\cdot)$ on $[0, \omega)$ yields $\psi_1(\cdot) \in L_\infty(0, \omega)$. As in part
(e) of the proof of Theorem 4.2.1 it follows that $\omega = \infty$. This proves the asser-
tions (i), (ii), and (iv), while (iii) follows from Lemma 2.1.8.
This completes the proof.   □

Remark 4.3.4
An analogous statement as in Theorem 4.2.4 for almost strictly positive real
system involving the feedback strategy $u(t) = -k(t)\Theta(t)y(t)$ is valid. The proof
is omitted and would be similar to the proof of Theorem 4.2.4 and 4.3.3.

4.4   Exponential stabilization via exponential weighting factor

The universal adaptive stabilizers presented in Sections 4.2 and 4.3 ensure that
the state $x(t)$ is decaying asymptotically to zero. That $x(t)$ decays exponentially
can only be guaranteed for scalar system of the form (1.1), with known positive
sign of the high-frequency gain, see the Introduction. For scalar systems with

unknown sign of the high-frequency gain, this does not hold true. To this end consider the scalar system

$$\begin{aligned} \dot{x}(t) &= -x(t) - u(t) \quad, \quad x(0) = 1 \\ y(t) &= x(t), \end{aligned}$$

which is stabilized by the feedback strategy

$$\begin{aligned} \dot{k}(t) &= y(t)^2 \qquad\qquad , \qquad k(0) = 0 \\ u(t) &= -k(t)K_{(Sok)(t)}y(t), \end{aligned}$$

where $\{K_1 = 1, K_2 = -1\}$ is a spectrum unmixing set for $\mathbb{R} \setminus \{0\}$, and a switching function is given by

$$S(\cdot) : \mathbb{R} \rightarrow \{1, 2\}, \; k \mapsto S(k) = \begin{cases} 1 & \text{if} \;\; k \in (-\infty, \tau_1) \\ i & \text{if} \;\; k \in [\eta_{N+i}, \eta_{N+i+1}) \\ & \qquad \text{for some} \;\; l \in \mathbb{N}_0, i \in \{1, 2\}. \end{cases}$$

If $\{\tau_i\}_{i \in \mathbb{N}}$ satisfies (4.5), then we are exactly in the position that the feedback (γ) in Theorem 4.2.1 yields stabilization. Specifying $\tau_1 = 1$, the closed-loop system, for all $t \geq 0$ such that $k(t) < \tau_1 = 1$, is

$$\begin{aligned} \dot{x}(t) &= [-1 + k(t)]x(t) \quad, \quad x(0) = 1 \\ \dot{k}(t) &= x(t)^2 \qquad\qquad , \qquad k(0) = 0. \end{aligned}$$

It is easily verified that the solution is given by

$$x(t) = \frac{1}{1+t} \quad \text{and} \quad k(t) = \frac{t}{1+t} .$$

Since $k(t) < 1$ for all $t \geq 0$, it follows that no switching occurs and the solution $x(t)$ decays to zero asymptotically, but not exponentially.

In this section, we shall design a universal adaptive stabilizer which ensures exponential decay of the state $x(t)$ of the closed-loop adaptive control system. The approach uses a time-invariant or time-varying (depending if we strengthen the minimum phase assumption or not) *exponential weighting factor* in the gain adaptation.

Many proofs are to a big extent similar to the ones of Section 4.2. To avoid repetition, we will only give the parts of proofs which differ from the one given in Section 4.2., and refer to parts from Section 4.2 whenever possible.

Definition 4.4.1
Let $\lambda > 0$. A system

$$\left. \begin{aligned} \dot{x}(t) &= Ax(t) + Bu(t) \\ y(t) &= Cx(t) + Du(t), \end{aligned} \right\} \tag{4.28}$$

where $(A, B, C, D) \in \mathbb{R}^{n \times n} \times \mathbb{R}^{n \times m} \times \mathbb{R}^{m \times n} \times \mathbb{R}^{m \times m}$, is called λ-*minimum phase*, if it satisfies the strengthened minimum phase condition

$$det \begin{bmatrix} sI_n - A & -B \\ -C & -D \end{bmatrix} \neq 0 \quad \text{for all} \quad s \in \mathbb{C} \quad \text{with} \quad \operatorname{Re} s \geq -\lambda.$$

The system (4.28) is called λ-*strictly positive real*, if it satisfies the Lur'e equations (3.6)-(3.8) for $\mu > \lambda$, and we say (4.28) is *almost λ-strictly positive real*, if there exists a $K \in \mathbb{R}^{m \times m}$, so that the feedback $u(t) = -K y(t) + r(t)$ yields a λ-strictly positive real system.

In order to show that (A, B, C, D) is λ-minimum phase if and only if $(A + \lambda I_n, B, C, D)$ is minimum phase, and an analogous statement for almost strict positive realness, we introduce the notation

$$v_\lambda(t) := e^{\lambda(t)t} v(t)$$

for any continuously differentiable $\lambda(\cdot) : \mathbb{R} \to \mathbb{R}$ and any vector-valued function $v(\cdot) : \mathbb{R} \to \mathbb{R}^r$.

For any $\lambda \geq 0$, a straightforward calculation yields that (4.28) is equivalent to

$$\left.\begin{aligned} \dot{x}_\lambda(t) &= [A + \lambda I_n] x_\lambda(t) &+ B u_\lambda(t) \\ y_\lambda(t) &= \qquad\quad C x_\lambda(t) &+ D u_\lambda(t). \end{aligned}\right\} \tag{4.29}$$

Proposition 4.4.2
Let $\lambda > 0$. The system (4.28) is λ-minimum phase (resp. almost λ-strictly positive real) if and only if the system (4.29) is minimum phase (resp. almost strictly positive real).

Proof: The statement concerning the minimum phase relationship follows from the definition. The relationship with respect to almost strict positive realness is a consequence of the following sequence of equivalent statements, where we use the notation of (3.14):

(4.28) is almost λ-strictly positive real

$\Leftrightarrow$ there exists a $K \in \mathbb{R}^{m \times m}$ so that $(\hat{A}, \hat{B}, \hat{C}, \hat{D})$ is λ-strictly positive real

$\Leftrightarrow$ there exist $P = P^T \in \mathbb{R}^{n \times n}$ positive-definite, $Q \in \mathbb{R}^{n \times m}$, $W \in \mathbb{R}^{m \times m}$, and $\mu > \lambda$ so that

$$
\begin{aligned}
P\hat{A} + \hat{A}^T P &= -QQ^T - 2\mu P \\
P\hat{B} &= \hat{C}^T - QW \\
W^T W &= \hat{D} + \hat{D}^T
\end{aligned}
$$

$\Leftrightarrow$ there exist $P = P^T \in \mathbb{R}^{n \times n}$ positive-definite, $Q \in \mathbb{R}^{n \times m}$, $W \in \mathbb{R}^{m \times m}$, and $\mu > \lambda$ so that

$$
\begin{aligned}
P(\hat{A} + \lambda I_n) + (\hat{A} + \lambda I_n)^T P &= -QQ^T - 2(\mu - \lambda)P \\
P\hat{B} &= \hat{C}^T - QW \\
W^T W &= \hat{D} + \hat{D}^T
\end{aligned}
$$

$\Leftrightarrow$ $(\hat{A} + \lambda I_n, \hat{B}, \hat{C}, \hat{D})$ is strictly positive real

$\Leftrightarrow$ $(A + \lambda I_n, B, C, D)$ is strictly positive real.

This completes the proof. $\square$

The equivalence between (4.26) and (4.28) already suggests how to modify the adaptation law $\dot{k}(t) = \|y(t)\|^p$ by introducing $\dot{k}(t) = e^{p\lambda t}\|y(t)\|^p$, and we obtain:

Theorem 4.4.3
Let $\lambda > 0$. If the following additional assumptions and modifications are applied to the following theorems

Theorem 4.2.1: (A, B, C) is λ-minimum phase, $\dot{k}(t) = \|y_\lambda(t)\|^p$,

Theorem 4.2.2: (A, B, C) is λ-minimum phase, $\frac{cA^2 b}{cAb} + 2\lambda < 0$, $\dot{k}(t) = \|y_\lambda(t)\|^2$,

Theorem 4.2.4: (A, B, C, D) is almost λ-strictly positive real,

Theorem 4.2.6: (A, B, C, D) is λ-minimum phase,

$$\dot{k}_1(t) = \|V^T y_\lambda(t)\|^2, \quad \dot{k}_2(t) = \|W^T y_\lambda(t)\|^2,$$

then the conclusions of these theorems remain valid and, in addition, $x(t)$ decays exponentially with decay rate at least λ.

Proof: The feedback laws (α)-(γ) in the theorems above can be rewritten by replacing $u(t)$ and $y(t)$ by $u_\lambda(t)$ and $y_\lambda(t)$, respectively. For the class of relative degree 2 systems considered in Theorem 4.2.2, note that, if $cb = 0$, then

$$c(A + \lambda I_n)b = cAb \quad \text{and} \quad \frac{c(A + \lambda I_n)^2 b}{cAb} = \frac{cA^2 b}{cAb} + 2\lambda,$$

and thus by the assumption and Theorem 2.2.6 it follows that $(A + \lambda I_n, b, c)$ is belonging to the class (4.11). Now it follows from Proposition 4.4.2 and the assumptions above that the system (4.29) satisfies the requirements of the theorems in terms of $(u_\lambda(t), x_\lambda(t), y_\lambda(t))$ instead of $(u(t), x(t), y(t))$. In all cases, we obtain, by the theorems listed above, $\lim_{t \to \infty} \|x_\lambda(t)\| = 0$, and hence $x(t)$ is decaying exponentially with decay rate at least λ. This proves the theorem. $\square$

We also can present a non-differentiable gain adaptation.

Theorem 4.4.4
Let $\lambda > 0$. If the following additional assumptions and modifications are applied to the following theorems

> Theorem 4.2.1: (A, B, C) is λ-minimum phase,
> $$k(t) = k_0 + \max_{s \in [0,t]} \|y_\lambda(t)\|^p,$$
>
> Theorem 4.2.2: (A, B, C) is λ-minimum phase, $\frac{cA^2b}{cAb} + 2\lambda < 0$,
> $$k(t) = k_0 + \max_{s \in [0,t]} \|y_\lambda(t)\|^2,$$
>
> Theorem 4.2.4: (A, B, C, D) is almost λ-strictly positive real,
>
> Theorem 4.2.6: (A, B, C, D) is λ-minimum phase,
> $$k_1(t) = k_0 + \max_{s \in [0,t]} \|Z^T y_\lambda(t)\|^2,$$
> $$k_2(t) = k_0 + \max_{s \in [0,t]} \|W^T y_\lambda(t)\|^2,$$

then the conclusions of the theorems remain valid and, in addition, $x(t)$ decays exponentially.

Proof: (a):   If $u(t) = -\mathcal{K}(t)y(t)$ stands for one of the feedback laws (α)-(γ) and is applied to (A, B, C, D), then the closed-loop system
$$\dot{x}(t) = \left[A - B\mathcal{K}(t)[I_m + \mathcal{K}(t)D]^{-1}C\right] x(t)$$
is well defined since it was shown in part (a) of the proof of Theorem 4.2.4 that $I_m + \mathcal{K}(t)D \in GL_m(\mathbb{R})$ for all $t \geq 0$. Since the value of $k(t)$ depends on past values of $y(s)$ on $[0, t]$, the above differential equation is a retarded functional differential equation. In all cases considered here, the differential equation fits into the framework studied in Chapter 2 in Hale (1977) and we obtain, in combination with the arguments used in previous proofs, a unique solution $x(\cdot) : [0, \omega) \to \mathbb{R}^n$, maximally extended over $[0, \omega)$, for some $\omega > 0$. The details are omitted.

(b):   In the previous theorems the hardest part of the proofs were to show that $k(\cdot) \in L_\infty(0, \omega)$. Since this does not involve the special structure of $k(\cdot)$, all theses proofs carry over and we have $k(\cdot) \in L_\infty(0, \omega)$.

(c):   Since $k(\cdot) \in L_\infty(0, \omega)$, it also follows as in the previous proofs, that $\omega = \infty$. Thus, we obtain for the finite limit $k_\infty := \lim_{t \to \infty} k(t)$,
$$e^{\lambda t} \|y(t)\| \leq \max_{s \in [0,t]} e^{\lambda s} \|y(s)\| \leq k_\infty - k_0,$$

and hence $y(t)$ decays exponentially.

If a minimum phase system of the form (2.3) is considered, then the second equation in (2.3) yields, in view of $\sigma(A_4) \subset \mathbb{C}_-$, that $z(t)$ is decaying exponentially, and hence $x(t)$ is as well.

If a strict positive real system of the form (3.13) is considered with $r(t) = -\mathcal{K}(t)y(t)$, then exponential decay of $x(t)$ follows since $\mathcal{K}(t)$ is bounded and $\hat{A}$ is exponentially stable.

This completes the proof. □

In Theorem 4.4.4 the knowledge of some $\lambda > 0$ is required. Since each minimum phase system is λ-minimum phase, and also each almost strictly positive real system is almost λ-strictly positive real, for some sufficiently small but unknown $\lambda > 0$, the aim is to find λ adaptively by using an exponential weighting factor tuned by $k(t)$. To this end we introduce the class of piecewise right differentiable functions $\lambda(\cdot) : [0, \infty) \to [0, \infty)$ with right continuous and locally Lipschitz derivative, where $\lambda(\cdot)$ satisfies the conditions

$$
\left.
\begin{aligned}
\lambda(k) \quad & \text{is non-increasing in } k \geq 0 \\
\lambda(k) \quad > \quad & 0 \text{ for all } k \geq 0 \\
\lim_{k \to \infty} \lambda(k) = \quad & 0.
\end{aligned}
\right\} \tag{4.30}
$$

A simple example is

$$
\lambda(k) = \begin{cases} 1 & \text{for } k \in [0, 1) \\ \frac{1}{1+k} & \text{for } k \geq 1. \end{cases}
$$

The assumption that the derivative of $\lambda(\cdot)$ is locally Lipschitz can be dropped at the expense of a non-unique solution of the closed-loop system.

If $\lambda(\cdot)$ belongs to (4.30), then a straightforward calculation yields that (4.28) is equivalent to

$$
\left.
\begin{aligned}
\dot{x}_{\lambda \circ k}(t) &= \left[A + \left[(\lambda \circ k)(t) + \tfrac{d}{dt}(\lambda \circ k)(t)t \right] I_n \right] x_{\lambda \circ k}(t) + Bu_{\lambda \circ k}(t) \\
y_{\lambda \circ k}(t) &= \qquad\qquad\qquad\qquad\qquad Cx_{\lambda \circ k}(t) + Du_{\lambda \circ k}(t).
\end{aligned}
\right\} \tag{4.31}
$$

If the gain adaptation is chosen to be

$$
\dot{k}(t) = e^{p(\lambda \circ k)(t)t} \|y(t)\|^p = \|y_{\lambda \circ k}(t)\|^p, \qquad k(0) \geq 0
$$

for $p \geq 1$, then the positive exponential $(\lambda \circ k)(t)$ will monotonically decrease as long as $k(t)$ is increasing. Finally, if the system does not stabilize before, $(\lambda \circ k)(t)$ will become smaller than some value $\lambda^* > 0$ so that the nominal system is λ^*-minimum phase or almost λ^*-strictly positive real, and exponential decay of $x(t)$ can be expected as in Theorem 4.4.3. This is the intuition behind the following Theorem 4.4.6.

A generalization of the inequality in Lemma 2.1.6, taking into account the exponential weighting factor, is needed for the proof of Theorem 4.4.6.

Lemma 4.4.5
Suppose the system (4.28) is minimum phase and $D = 0$, $det(CB) \neq 0$, and $\lambda(\cdot)$ belongs to (4.28). If any locally integrable $u(\cdot) : [0, \omega) \to \mathbb{R}^m$, with $\omega \in (0, \infty]$, is applied to the system (4.28), then for every initial condition $x(0) = x_0 \in \mathbb{R}^n$ and positive-definite matrix $P = P^T \in \mathbb{R}^{m \times m}$, there exists $M > 0$ (depending only on A, B, C and P) such that, for all $t \in [0, \omega)$,

$$\frac{1}{2}\|y_\lambda(t)\|_P^2 \leq M\|x(0)\|^2 + M \int_0^t \|y_\lambda(s)\|^2 ds + \int_0^t \langle y_\lambda(s), P C B u_\lambda(s)\rangle ds$$

for all $t \geq 0$.

Proof: Supposing that (4.28) is of the form (2.3), we obtain, for all $s \in [0, \omega)$,

$$\frac{1}{2}\frac{d}{ds}\|y_\lambda(s)\|_P^2 \leq \left\langle y_\lambda(s), P[e^{\lambda(s)s}\dot{y}(s) + (\lambda(s) + \dot{\lambda}(s)s)y_\lambda(s)]\right\rangle$$

$$= [\lambda(s) + \dot{\lambda}(s)s]\|y_\lambda(s)\|_P^2 + e^{\lambda(s)s}\langle y_\lambda(s), P\dot{y}(s)\rangle$$

$$= [\lambda(s) + \dot{\lambda}(s)s]\|y_\lambda(s)\|_P^2$$
$$\quad + \langle y_\lambda(s), P A_1 y_\lambda(s) + P A_2 z_\lambda(s) + P C B u_\lambda(s)\rangle .$$

$z_\lambda(t)$ satisfies

$$\dot{z}_\lambda(t) = A_3 y_\lambda(t) + [A_4 + (\lambda(t) + \dot{\lambda}(t)t)I_{n-m}]z_\lambda(t).$$

Since $\dot{\lambda}(t) \leq 0$, $\lim_{t \to \infty} \lambda(t) = 0$, and $\sigma(A_4) \subset \mathbb{C}_-$, there exist $M_1, \hat{\omega} > 0$ such that

$$\left\|e^{\int_0^t [A_4 + (\lambda(s) + \dot{\lambda}(s)s)I_{n-m}]ds}\right\| \leq M_1 e^{-\hat{\omega}t} \quad \text{for all} \quad t \in [0, \omega).$$

Therefore, cf. e.g. Vidyasagar pp. 250-254, the operator

$$\mathcal{L}: L_p(0, \omega) \to L_p(0, \omega)$$

$$y(\cdot) \mapsto \left(t \mapsto A_2 \int_0^t e^{\int_s^t [A_4 + (\lambda(\mu) + \dot{\lambda}(\mu)\mu)I_{n-m}]d\mu} A_3 y(s) ds\right)$$

is well defined and bounded for all $p \in [1, \infty]$. Defining

$$w_\lambda(t) = e^{\int_0^t [A_4 + (\lambda(\mu) + \dot{\lambda}(\mu)\mu)I_{n-m}]d\mu} x_\lambda(0),$$

we obtain

$$\|z_\lambda(s)\| \leq \|w_\lambda(s)\| + \|\mathcal{L}(y_\lambda)(s)\|,$$

and hence, for all $s \in [0,\omega)$,

$$\frac{1}{2}\frac{d}{ds}\|y_\lambda(s)\|_P^2 \leq [\lambda(s)\|P\| + \|PA_1\|] \|y_\lambda(s)\|$$

$$+ \|PA_2\|\|y_\lambda(s)\| [\|w_\lambda(s)\| + \|\mathcal{L}(y_\lambda)(s)\|]$$

$$+ \langle y_\lambda(s), +PCBu_\lambda(s)\rangle.$$

This inequality is analog to (2.8) and the remainder of the proof follows in a similar manner as in the proof of Lemma 2.1.6. □

Theorem 4.4.6
Suppose $\lambda(\cdot) : [0,\infty) \to [0,\infty)$ is not identical zero and belongs to (4.30). If the gain adaptation is replaced in the following theorems by:

$$\text{Theorem 4.2.1 and 4.2.4:} \quad \dot{k}(t) = \|y_{\lambda o k}(t)\|^2, \qquad k(0) \geq 0$$

$$\text{Theorem 4.2.6:} \quad \dot{k}_1(t) = \|Z^T y_{\lambda o k}(t)\|^2, \quad k_1(0) \geq 0,$$

$$\dot{k}_2(t) = \|W^T y_{\lambda o k}(t)\|^2, \quad k_2(0) \geq 0,$$

then the conclusions of the theorems remain valid and, moreover, each solution $x(t)$ of the closed-loop system decays exponentially.

Proof: By the assumptions on $\lambda(\cdot)$, it follows in a similar manner as in the proof of theorems listed above, that for every $(x_0, k(0)) \in \mathbb{R}^{n+1}$, resp. $(x_0, k_1(0), k_2(0)) \in \mathbb{R}^{n+2}$, the closed-loop system has a unique solution $(x(\cdot), k(\cdot)) : [0,\omega) \to \mathbb{R}^{n+1}$, resp. $(x(\cdot), k_1(\cdot), k_2(\cdot)) : [0,\omega) \to \mathbb{R}^{n+2}$, maximally extended over $[0,\omega)$ for some $\omega > 0$.

(a): Suppose feedback of the form $u(t) = -\mathcal{K}(t)y(t)$ is applied in Theorem 4.2.4. Let $K \in \mathbb{R}^{m \times m}$ so that (3.13) is strictly positive real and (3.6)-(3.8) hold for the system matrices (3.14). Hence the derivative of $V(x_{\lambda o k}) = \langle x_{\lambda o k}, Px_{\lambda o k}\rangle$ along (3.15), as in the proof of Lemma 3.1.6, is

$$\frac{d}{dt}V(x_{\lambda o k}(t)) = e^{2(\lambda o k)(t)t} \langle \dot{x}(t), Px_{\lambda o k}(t)\rangle$$

$$+ 2\left[\frac{d}{dt}(\lambda \circ k)(t)t + (\lambda \circ k)(t)\right] V(x_{\lambda o k}(t))$$

$$\leq -2\left[\mu - \frac{d}{dt}(\lambda \circ k)(t)t - (\lambda \circ k)(t)\right] V(x_{\lambda o k}(t))$$

$$+ 2\|K\|\|y_{\lambda o k}(t)\|^2 - \langle y_{\lambda o k}(t), [\mathcal{K}(t) + \mathcal{K}^T(t)]y_{\lambda o k}(t)\rangle$$

for some $\mu > 0$, $K \in \mathbb{R}^{m \times m}$, and all $t \in [0, \omega)$. If $k(\cdot) \notin L_\infty(0, \omega)$, then, by (4.30) and since $\frac{d}{dt}(\lambda \circ k)(t) \leq 0$, there exists a $t_0 \in (0, \omega)$ such that

$$\mu - \frac{d}{dt}(\lambda \circ k)(t)t - (\lambda \circ k)(t) > 0 \quad \text{for all} \quad t \in [t_0, \omega).$$

Hence integration of $\frac{d}{ds}V(x_{\lambda \circ k}(s))$ over $[t_0, t]$ yields

$$V(x_{\lambda \circ k}(t)) \leq V(x_{\lambda \circ k}(t_0)) + \int_{t_0}^t \langle y_{\lambda \circ k}(\tau), [\mathcal{K}(\tau) + \mathcal{K}^T(\tau)]y_{\lambda \circ k}(\tau) \rangle \, d\tau$$
$$+ 2\|K\|[k(t) - k(t_0)]$$

Using similar arguments as in parts (a)-(d) resp. (e) of the proof of Theorem 4.2.4, we conclude $k(\cdot) \in L_\infty(0, \omega)$, and hence $y_{\lambda \circ k}(\cdot) \in L_2(0, \omega)$, whence $\omega = \infty$. This proves the assertions (i), (ii), and (iv) for the modified Theorem 4.2.4.

(b): If

$$\lambda_\infty := \lim_{t \to \infty}(\lambda \circ k)(t),$$

it follows that

$$y_\zeta(\cdot) \in L_2(0, \omega) \quad \text{for all} \quad \zeta \in (0, \lambda_\infty).$$

If (A, B, C) is supposed to be of the form (2.3), then a straightforward calculation yields that it can be written as

$$\dot{y}_\zeta(t) = [A_1 + \zeta I_m]y_\zeta(t) + A_2 z_\zeta(t) + CBu_\zeta(t)$$
$$\dot{z}_\zeta(t) = A_3 y_\zeta(t) + [A_4 + \zeta I_{n-m}]z_\zeta(t)$$

with $\sigma(A_4) \subset \mathbb{C}_-$. Therefore, for $\zeta > 0$ sufficiently small, Lemma 2.1.8 yields $\lim_{t \to \infty} x_\zeta(t) = 0$. This proves assertion (iii) and exponential decay of $x(t)$.

(c):   We omit the proof for the modified Theorem 4.2.4 and 4.2.6. It can be carried out by using similar ideas as in (a), (b) above, and as in the original proof of Theorem 4.2.1 and 4.2.6., by using the inequality in Lemma 4.4.5. This completes the proof.   □

Remark 4.4.7
Exponential stabilization is also possible by using the idea of exponential weighting as in the Theorem 4.4.6 in combination with the switching strategy based on a switching decision function, see Section 4.3. This has been proved in Ilchmann and Owens (1991a).

4.5 Exponential stabilization via piecewise constant gain

The disadvantage of the feedback strategy introduced in Section 4.4 is that the gain adaptation is unbounded. To overcome this, we introduce a piecewise constant gain adaptation tuned by the function $s(t) = \int_0^t \|y(\tau)\|^2 d\tau$ and a pre-specified *sequence of thresholds* $T_0 < T_1 < \ldots$ with $\lim_{i \to \infty} T_i = \infty$. In case of $\sigma(CB) \subset \mathbb{C}_+$ or $\mathbb{C}_+$, the idea is as follows: At each time t_i when the 'tuning function' $s(t) = \int_0^t \|y(s)\|^2 ds + T_0$ reaches a threshold T_i, the feedback law will be changed to $k(t) = (-1)^i T_i y(t)$. The benefit of this modified adaptive strategy is, that the part $\dot{x}(t) = \left[A - (-1)^{i+1} T_{i+1} BC\right] x(t)$ of the closed-loop system is a time-invariant system on intervals $[t_i, t_{i+1})$. As long as $s(t)$ reaches the next threshold T_{i+1}, a higher gain with opposite sign is implemented until finally, the gain is large enough and the sign is correct so that $x(t)$ converges to zero. Thus the switching algorithm will stop, and we end up with a constant system and consequently, each asymptotic decaying solution of the system is exponentially decaying. We will show that this idea can be extended to strictly proper minimum phase systems with $|CB| \neq 0$, to single-input, single-output, positive or negative high-gain stabilizable systems, and to certain proper minimum phase systems.

Apart from exponential stabilization, the other advantage of this switching strategy is that for single-input, single-output systems the set of sequences of thresholds which yield an exponentially stable terminal system, has nice topological properties, see Section 8.2.

Theorem 4.5.1

Suppose

$$\left.\begin{array}{rcll} \dot{x}(t) & = & Ax(t) + Bu(t) & , \quad x(0) = x_0 \\ y(t) & = & Cx(t), \end{array}\right\} \tag{4.32}$$

with $(A, B, C) \in \mathbb{R}^{n \times n} \times \mathbb{R}^{n \times m} \times \mathbb{R}^{m \times n}$, is minimum phase, $p \geq 1$, and $0 \leq T_0 < T_1 < \ldots$ is a strictly increasing sequence of real numbers. Let $\{K_1, \ldots, K_N\}$ denote a spectrum unmixing set for $GL_m(\mathbb{R})$. If the adaptation law

$$\begin{array}{rcll} \dot{s}(t) & = & \|y(t)\|^p & , \quad s(0) = T_0 \\ k(t) & = & T_i, & , \quad \text{if} \quad s(t) \in [T_{i-1}, T_i) \end{array}$$

together with one of the following feedback laws (and additional assumptions on the growth rate of $\{T_i\}_{i\in\mathbb{N}}$)

$(\alpha):$ $u(t) = -k(t)y(t)$ if $\sigma(CB) \subset \mathbb{C}_+$

and $\lim_{i\to\infty}[T_i - T_{i-1}] = \infty,$

$(\beta):$ $u(t) = -N(s(t))y(t)$ if $\sigma(CB) \subset \mathbb{C}_+$ or $\mathbb{C}_-$ and $p = 2$

$N(s(t)) = (-1)^i T_i$ if $s(t) \in [T_{i-1}, T_i)$

and

$$\inf_{l\in\mathbb{N}} \frac{\sum_{i=1}^{l}(-1)^i T_i(T_i-T_{i-1})}{T_l - T_0} = -\infty, \quad \sup_{l\in\mathbb{N}} \frac{\sum_{i=1}^{l}(-1)^i T_i(T_i-T_{i-1})}{T_l - T_0} = +\infty,$$

$(\gamma):$ $u(t) = -k(t)K(t)y(t)$ if $|CB| \neq 0$

$K(t) = K_{i\,mod\,N}$ if $s(t) \in [T_{i-1}, T_i)$

and $\lim_{i\to\infty} \frac{T_i}{T_{i+1}} = 0,$

and arbitrary $x_0 \in \mathbb{R}^n$, is applied to (4.32), then the closed-loop system has the properties

(i) the unique solution $(x(\cdot), s(\cdot)) : [0, \infty) \to \mathbb{R}^{n+1}$ exists,

(ii) $\lim_{t\to\infty} s(t) = s_\infty$ exists and there exist only finitely many switches,

(iii) $x(t)$ decays exponentially as t tends to ∞.

Proof: (a):   Discontinuities of the right hand side of the closed-loop system occur whenever $s(t)$ leaves an interval $[T_{i-1}, T_i)$, that is at times

$$t_i := \inf\{t \geq t_{i-1} \,|\, s(t) = T_i\}, \qquad i = 1, 2, \ldots \qquad , t_0 := 0.$$

There exists a unique solution on the interval $[0, t_1)$, and, if $t_1 < \infty$, then the finite limit $x(t_1) := \lim_{t\to t_1} x(t)$ exists. Proceeding in this way for the next intervals, it follows that the solution $x(t)$ of the closed-loop system exists on some interval $[0, \omega)$, where $\lim_{i\to\infty} t_i = \omega \in (0, \infty]$.

(b):   Consider the feedback $u(t) = -k(t)K(t)y(t)$ described in (γ). We shall show that $s(\cdot) \in L_\infty(0, \omega)$. Suppose the contrary, i.e. infinitely many switches occur. By assumption, there exists $K_i \in \{K_1, \ldots, K_N\}$ such that $\sigma(CBK_i) \subset \mathbb{C}_+$. Let $P = P^T \in \mathbb{R}^{m\times m}$ be the unique positive-definite solution of

$$K_i^T(CB)^T P + PCBK_i = 2I_m \qquad (4.33)$$

and choose $\alpha > 0$ so that

$$-K_l^T(CB)^T P - PCBK_l \leq 2\alpha I_m \quad \text{for all} \quad l \in \underline{N}. \qquad (4.34)$$

Without restriction of generality one may assume that $i = N$, otherwise start at a different time t_j with initial condition $x(t_j)$. In order to apply the inequality in Lemma 2.1.6, we calculate, by using (4.33), (4.34), and $\dot{s}(t) = \|y(t)\|^p$

$$
-\int_{t_{jN}}^{t_{(j+1)N}} k(\tau)\|y(\tau)\|_P^{p-1}\langle\beta(y(\tau)), P\,CBK(\tau)y(\tau)\rangle d\tau
$$

$$
\leq \|P\|^{\frac{p-2}{2}}\alpha \int_{t_{jN}}^{t_{(j+1)N-1}} k(\tau)\|y(\tau)\|^p\,d\tau - \mu_{\min}(P)^{\frac{p-2}{2}} \int_{t_{(j+1)N-1}}^{t_{(j+1)N}} k(\tau)\|y(\tau)\|^p\,d\tau
$$

$$
= \alpha' T_{(j+1)N-1}[T_{(j+1)N-1} - T_{jN}] - \beta' T_{(j+1)N}[T_{(j+1)N} - T_{(j+1)N-1}]
$$

$$
= T_{(j+1)N}^2 \left[\alpha'\frac{T_{(j+1)N-1}^2 - T_{(j+1)N-1}T_{jN}}{T_{(j+1)N}^2} + \beta'\frac{T_{(j+1)N-1}}{T_{(j+1)N}} - \beta'\right]
$$

for

$$
\alpha' := \alpha\|P\|^{\frac{p-2}{2}}, \qquad \beta' := \mu_{\min}(P)^{\frac{p-2}{2}}.
$$

Since $\lim_{i\to\infty}\frac{T_i}{T_{i+1}} = 0$, the right hand side of the above inequality tends to $-\beta' T_{(j+1)N}^2$ as $j \to \infty$. Thus there exists a $M_1 > 0$ such that

$$
-\sum_{j=0}^{l-2} \int_{t_{jN}}^{t_{(j+1)N}} k(\tau)\|y(\tau)\|_P^{p-1}\langle\beta(y(\tau)), P\,CBK(\tau)y(\tau)\rangle d\tau \leq M_1 \text{ for all } l \in \mathbb{N}.
$$

$$(4.35)$$

We also have

$$
\frac{-1}{T_{lN} - T_0} \int_{t_{(l-1)N}}^{t_{lN}} k(\tau)\|y(\tau)\|_P^{p-1}\langle\beta(y(\tau)), P\,CBK(\tau)y(\tau)\rangle d\tau
$$

$$
\leq \frac{T_{lN}^2}{T_{lN} - T_0}\left[\alpha'\frac{T_{lN-1}^2 - T_{lN-1}T_{jN}}{T_{lN}^2} + \beta'\frac{T_{lN-1}}{T_{lN}} - \beta'\right] \quad (4.36)
$$

and the right hand side of (4.36) tends to $-\beta' T_{lN}$ as $l \to \infty$.
It follows from Lemma 2.1.6 that for all $l \in \mathbb{N}$ and for some $M > 0$, we have

$$
\frac{1}{p}\|y(t_{lN})\|_P^p \leq M + M(T_{lN} - T_0)
$$

$$
-\sum_{j=0}^{l-1} \int_{t_{jN}}^{t_{(j+1)N}} k(\tau)\|y(\tau)\|_P^{p-1}\langle\beta(y(\tau)), P\,CBK(\tau)y(\tau)\rangle d\tau
$$

$$
= M + (T_{lN} - T_0)\,[M-
$$

$$
\frac{1}{T_{lN} - T_0}\sum_{j=0}^{l-1} \int_{t_{jN}}^{t_{(j+1)N}} k(\tau)\|y(\tau)\|_P^{p-1}\langle\beta(y(\tau)), P\,CBK(\tau)y(\tau)\rangle d\tau\Bigg] \quad (4.37)
$$

Inserting (4.35) and (4.36) into (4.37) yields

$$\frac{1}{p}\|y(t_{lN})\|_P^p \leq M + (T_{lN} - T_0)\left[M + \frac{M_1}{T_{lN} - T_0}\right.$$
$$\left.\frac{T_{lN}^2}{T_{lN} - T_0}\left(\alpha'\frac{T_{lN-1}^2 - T_{lN-1}T_{jN}}{T_{lN}^2} + \beta'\frac{T_{lN-1}}{T_{lN}} - \beta'\right)\right].$$

The right hand side of the above inequality tends to $-\infty$ as $l \to \infty$. This contradicts the non-negativeness of the left hand side, and proves $s(\cdot) \in L_\infty(0, \omega)$.

(c):   Since $k(t)$ in the right hand side of the closed-loop system is bounded, it follows from classical theory of differential equations that $\omega = \infty$. This proves the assertions (i) and (ii) for the feedback (γ).
To prove exponential decay of $x(t)$, let $t^* > 0$ such that

$$K(t) = T_M K_{M \bmod N} \quad \text{for all} \quad t \geq t^*.$$

By Lemma 2.1.3, the terminal system

$$\dot{x}(t) = [A + T_M B K_{M \bmod N} C]x(t), \qquad x(0) = x(t_{M-1})$$

can be expressed as

$$\begin{aligned}\dot{y}(t) &= (A_1 + T_M K_{M \bmod N} CB)y(t) &+& A_2 z(t)\\ \dot{z}(t) &= & A_3 y(t) &+& A_4 z(t).\end{aligned}$$

Since $y(\cdot) \in L_p(0, \infty)$, Lemma 2.1.8 yields $\lim_{t \to \infty} x(t) = 0$ and hence, in view of the time-invariance and linearity of the above system, $x(t)$ decays exponentially.

(d):   Consider the feedback (β). Let $P = P^T \in \mathbb{R}^{m \times m}$ be the unique positive-definite solution of

$$PCB + (CB)^T P = 2\tau I_m.$$

for some $\tau \in \{+1, -1\}$. Lemma 2.1.6 yields

$$\frac{1}{2}\|y(t_l)\|_P^2 \leq M + M\int_0^{t_l}\|y(\mu)\|^2 d\mu - \tau\int_0^{t_l} N(s(\mu))\|y(\mu)\|^2 d\mu$$

$$= M + M(T_l - T_0) - \tau\sum_{i=0}^{l}(-1)^i T_i[T_i - T_{i-1}]$$

$$= M + (T_l - T_0)\left[M - \tau\frac{\sum_{i=0}^{l}(-1)^i T_i(T_i - T_{i-1})}{T_l - T_0}\right]$$

for some $M > 0$. If $s(\cdot) \notin L_\infty(0, \omega)$, then the right hand side of the above inequality takes, by the growth rate assumption on $\{T_i\}_{i \in \mathbb{N}}$, negative values.

This is a contradiction, and hence $s(\cdot) \in L_\infty(0, \omega)$.
The remainder of the proof is similar as in (c), it is omitted.

(e): The proof for the feedback (α) is a simplification of (d), it is omitted.
This completes the proof of the theorem.   □

Remark 4.5.2

(i) If the feedback (β) in Theorem 4.5.1 is applied, then the growth condition on $\{T_i\}_{i \in \mathbb{N}}$, is a discrete analogy to $N(\cdot)$ being a Nussbaum function. If $p \neq 2$, then, in the continuous feedback case in Theorem 4.2.1, we had to assume that $N(\cdot)$ is scaling-invariant. This can be done in a discrete context as well and a sequence $\{T_i\}_{i \in \mathbb{N}} \subset \mathbb{R}_+$ is called *scaling-invariant* if, for every $\alpha, \beta > 0$, it satisfies

$$\sup_{r \in \mathbb{N}} \frac{1}{T_{2r+1} - T_0} \left[\alpha \sum_{i=0}^{r} T_{2i+1}[T_{2i+1} - T_{2i}] - \beta \sum_{i=1}^{r} T_{2i}[T_{2i} - T_{2i-1}] \right] = +\infty,$$

$$\inf_{r \in \mathbb{N}} \frac{1}{T_{2(r+1)} - T_0} \left[\alpha \sum_{i=0}^{r} T_{2i+1}[T_{2i+1} - T_{2i}] - \beta \sum_{i=1}^{r+1} T_{2i}[T_{2i} - T_{2i-1}] \right] = -\infty.$$

(ii) It is easy to see that the sequence $T_i := i^2$ satisfies the growth condition in (β) of Theorem 4.5.1 but not the one in (γ). i^2 is also not scaling-invariant.
The sequence $T_i = (-1)^i T_{i-1}^2$, $T_0 > 1$ satisfies (β) and (γ) and is scaling-invariant.

(iii) If for $p \neq 2$ the switching strategy in Theorem 4.5.1 is modified to $\dot{s}(t) = \|y(t)\|^p$, and $\{T_i\}_{i \in \mathbb{N}}$ is assumed to be scaling-invariant, then the same result is valid. This can easily be seen since we have, by using the notation as in part (d) of the proof and (2.7),

$$\frac{1}{p}\|y(t)\|_p^p \leq M + M[T_l - T_0] - \int_0^{t_l} \tilde{N}(s(\mu))\dot{s}(\mu)d\mu$$

where

$$\tilde{N}(s) := \begin{cases} \tau N(s)\|P\|^{\frac{p-2}{2}} & , \text{ if } \tau N(s) \leq 0 \\ \tau N(s)\mu_{\min}(P)^{\frac{p-2}{2}} & , \text{ if } \tau N(s) > 0. \end{cases}$$

If $s(\cdot) \notin L_\infty(0, \omega)$, then integration of the above inequality yields, by the scaling-invariance property, that the right hand side takes negative values, which is a contradiction. All the other arguments are as before and the statements of Theorem 4.5.1 follow.

We shall show that all positive or negative high-gain stabilizable single-input single-output systems, i.e. the class (4.11), can be adaptively stabilized by piecewise constant gain implementation, having the benefit that the solution of the closed-loop system decays exponentially.

Theorem 4.5.3

If

$$
\begin{aligned}
\dot{x}(t) &= Ax(t) + bu(t) \quad , \quad x(0) = x_0 \\
y(t) &= cx(t),
\end{aligned}
\right\}
\tag{4.38}
$$

$(A, b, c) \in \mathbb{R}^{n \times n} \times \mathbb{R}^n \times \mathbb{R}^{1 \times n}$, is an arbitrary realization of a transfer function belonging to (4.11), and $\{T_i\}_{i \in \mathbb{N}}$ is defined by $T_i = T_{i-1}^{2+\varepsilon}$, with $T_0 > 1$, $\varepsilon > 0$, then the feedback strategy

$$
\begin{aligned}
u(t) &= -N(s(t))y(t) \\
N(s(t)) &= (-1)^i \sqrt{T_i} \quad \text{if} \quad s(t) \in [T_{i-1}, T_i) \\
\dot{s}(t) &= y(t)^2 \quad , \quad s(0) = T_0
\end{aligned}
\right\}
$$

applied to (4.38), for arbitrary $x_0 \in \mathbb{R}^n$, yields a closed-loop system with the properties

(i) the unique solution $(x(\cdot), s(\cdot)) : [0, \infty) \to \mathbb{R}^{n+1}$ exists,

(ii) $\lim_{t \to \infty} s(t) = s_\infty$ exists and there exist only finitely many switches,

(iii) $x(t)$ decays exponentially as t tends to ∞.

Proof: (a): Define

$$
t_i := \inf \{t > t_{i-1} \mid s(t) = T_i\} \quad \text{for} \quad i = 1, 2, \ldots, \quad t_0 := 0.
$$

There exists a unique solution on the interval $[0, t_1)$, and, if $t_1 < \infty$, then the finite limit $x(t_1) := \lim_{t \to t_1} x(t)$ exists. Proceeding in this way for the next intervals, it follows that the solution $x(t)$ of the closed-loop system exists on some interval $[0, \omega)$, where $\lim_{i \to \infty} t_i = \omega \in (0, \infty]$.

(b): If $cb \neq 0$, then the result follows from Theorem 4.5.1. Note that $\{T_i\}_{i \in \mathbb{N}}$ satisfies the growth condition required in Theorem 4.5.1 for the feedback (β).

(c) It remains to consider the case when (A, b, c) is of relative degree 2. Suppose $s(\cdot) \notin L_\infty(0, \omega)$. Let $i_0 \in \mathbb{N}$ so that

$$
cAb(-1)^{i_0-1}T_{i_0} > M
$$

with M satisfying the requirements of Lemma 2.2.9. It then follows that for every

$$
u(t) = (-1)^{i-1}T_i \, y(t) \quad \text{with} \quad i \in i_0 + 2\mathbb{N}
$$

we have

$$y(t)^2 \leq \hat{M} e^{-\hat{\lambda}(t-t_{i-1})} \|x(t_{i-1})\|^2 \quad \text{for all} \quad t \in [t_{i-1}, t_i)$$

for some $\hat{M}, \hat{\lambda} > 0$. Therefore,

$$T_i - T_{i-1} = \int_{t_{i-1}}^{t_i} y(s)^2 ds \leq \frac{\hat{M}}{\hat{\lambda}} \|x(t_{i-1})\|^2 \quad \text{for all} \quad i \in i_0 + 2\mathbb{N} \qquad (4.39)$$

Choose $T^* \in \mathbb{R}$ so that

$$\sigma(A^*) \subset \mathbb{C}_- \quad \text{where} \quad A^* := A - T^* bc,$$

and the solution of

$$\dot{x}(t) = Ax(t) + bu(t) = A^* x(t) - [N(s(t)) - T^*]by(t), \qquad x(0) \in \mathbb{R}^n$$

satisfies, by exponential stability of A^*,

$$\|x(t)\| \leq L e^{-\lambda t} \|x(0)\| + L\|b\| \left[|N(s(\cdot))|_{L_\infty(0,t)} + |T^*| \right] \int_0^t e^{-\lambda(t-s)} |y(s)| ds.$$

for some $L, \lambda > 0$. This yields, by using (4.17),"

$$\|x(t)\| \quad \leq \quad L\|x(0)\| + \frac{L\|b\|}{\sqrt{\lambda}} \left[|N(s(\cdot))|_{L_\infty(0,t)} + |T^*| \right] \sqrt{\int_0^{t_i} y(s)^2 ds},$$

and hence by taking square roots in (4.39),

$$\sqrt{T_i - T_{i-1}} \leq M_1 + M_2 \left[\sqrt{T_{i-1}} + |N^*| \right] \sqrt{T_{i-1}}$$

where

$$M_1 := \sqrt{\frac{\hat{M}}{\hat{\lambda}}} L\|x(0)\|, \qquad M_2 := \sqrt{\frac{\hat{M}}{\hat{\lambda}}} \frac{L\|b\|}{\sqrt{\lambda}}.$$

The above inequality is equivalent to

$$1 \quad \leq \quad \frac{M_1}{\sqrt{T_i - T_{i-1}}} + M_2 \left[\sqrt{T_{i-1}} + |N^*| \right] \sqrt{\frac{T_{i-1}}{T_i - T_{i-1}}}.$$

Since

$$\lim_{i \to \infty} \frac{T_{i-1}^2}{T_i - T_{i-1}} = \lim_{i \to \infty} \frac{1}{T_{i-1}^\epsilon - T_{i-1}^{-1}} = 0,$$

it follows that the right hand side of the above inequality tends to 0 as $i \to \infty$, contradicting to be greater or equal to 1. This proves $s(\cdot) \in L_\infty(0, \omega)$, and hence $y(\cdot) \in L_2(0, \omega)$ and $N(s(\cdot)) \in L_\infty(0, \omega)$.

(d):   The remainder of the proof follows from Lemma 2.1.8 and by similar arguments as in the proof of part (c) of Theorem 4.5.1, it is omitted.   □

Remark 4.5.4
The growth of the sequence $\{T_i\}_{i\in\mathbb{N}}$ and the magnitude of $N(s(t))$ in the feedback strategy in Theorem 4.5.3 can be generalized as follows. Suppose $0 \le T_1 < T_2 < \ldots$ is strictly increasing and $N(s(t))$ is defined for some $h(\cdot) : [0,\infty) \to [0,\infty)$ by

$$N(s(t)) = h(T_i) \quad \text{if} \quad s(t) \in [T_{i-1}, T_i).$$

If $h(\cdot)$ satisfies

$$\lim_{i\to\infty} \frac{h(T_{i-1})T_{i-1}}{T_i - T_{i-1}} = 0,$$

and the assumption of Theorem 4.5.3 are fulfilled, then the same result is valid. This can be seen directly from part (c) of the proof.

Next we will show that piecewise constant gain implementation stabilizes also proper, not necessarily strictly proper, systems.

Theorem 4.5.5
Suppose $G(\cdot) \in GL_m(\mathbb{R}(s))$ is minimum phase with minimal realization

$$\left.\begin{array}{rcl} \dot{x}(t) & = & Ax(t) + Bu(t) \\ y(t) & = & Cx(t) + Du(t), \end{array} \quad , \quad x(0) = x_0 \right\} \tag{4.40}$$

where $(A, B, C, D) \in \mathbb{R}^{n\times n} \times \mathbb{R}^{n\times m} \times \mathbb{R}^{m\times n} \times \mathbb{R}^{m\times m}$, and $\text{rk}B = m \le n$. Let $\{K_1, \ldots, K_N\}$ denote a spectrum unmixing set for $GL_{m-r}(\mathbb{R})$.
Suppose furthermore, (4.40) satisfies (I)-(III) below:

(I) There exists an orthogonal matrix $[Z, W]$, $Z \in \mathbb{R}^{m\times(m-r)}$, $W \in \mathbb{R}^{m\times r}$, $r = \text{rk}\, D$ such that

$$[Z, W]^T D[Z, W] = \begin{bmatrix} 0, & 0 \\ 0, & W^T DW \end{bmatrix} \tag{4.41}$$

(II) $\det(Z^T CBZ) \ne 0$,

(III) $W^T[D + D^T]W \ge 0$.

If one of the feedback strategies

$(\alpha):$ $u(t) = -k(t)y(t)$, if $\sigma(Z^T CBZ) \subset \mathbb{C}_+$
$\quad\quad\; \dot{s}(t) = \|y(t)\|^2$, $s(0) = T_0$
$\quad\quad\; k(t) = T_i$, if $s(t) \in [T_{i-1}, T_i)$
$\quad\quad\;$ and $\{T_i\}_{i \in \mathbb{N}}$ satisfies $\lim_{i \to \infty} [T_{i+1} - T_i] = \infty,$

$(\beta):$ $u(t) = -[Z, W] \begin{bmatrix} N(s_1(t))I_{m-r}, & 0 \\ 0 & k_2(t)I_r \end{bmatrix} [Z, W]^T y(t),$
$\quad\quad\quad\quad\quad\quad\quad\quad\quad\quad\quad\quad\quad$ if $\sigma(Z^T CBZ) \subset \mathbb{C}_+$ or $\mathbb{C}_-$

$\quad\quad\; \dot{s}_1(t) = \|Z^T y(t)\|^2$, $s_1(0) = T_0$
$\quad\quad\; N(s_1(t)) = (-1)^i T_i$, if $s_1(t) \in [T_{i-1}, T_i)$
$\quad\quad\; \dot{s}_2(t) = \|W^T y(t)\|^2$, $s_2(0) = T_0$
$\quad\quad\; k_2(t) = T_i$, if $s_2(t) \in [T_{i-1}, T_i)$
$\quad\quad\;$ and $\{T_i\}_{i \in \mathbb{N}}$ satisfies

$$\inf_{l \in \mathbb{N}} \frac{\sum\limits_{i=1}^{l}(-1)^i T_i (T_i - T_{i-1})}{T_l - T_0} = -\infty, \quad \sup_{l \in \mathbb{N}} \frac{\sum\limits_{i=1}^{l}(-1)^i T_i (T_i - T_{i-1})}{T_l - T_0} = +\infty,$$

$(\gamma):$ $u(t) = -[Z, W] \begin{bmatrix} k_1(t)K(t), & 0 \\ 0 & k_2(t)I_r \end{bmatrix} [Z, W]^T y(t)$ if $|Z^T CBZ| \neq 0$

$\quad\quad\; \dot{s}_1(t) = \|Z^T y(t)\|^2$, $s_1(0) = T_0$
$\quad\quad\; K(t) = K_{i \bmod N}$, if $s_1(t) \in [T_{i-1}, T_i)$
$\quad\quad\; \dot{s}_2(t) = \|W^T y(t)\|^2$, $s_2(0) = T_0$
$\quad\quad\; k_2(t) = T_i$, if $s_2(t) \in [T_{i-1}, T_i)$
$\quad\quad\;$ and $\{T_i\}_{i \in \mathbb{N}}$ satisfies $\lim_{i \to \infty} \frac{T_i}{T_{i+1}} = 0,$

then, for arbitrary $x_0 \in \mathbb{R}^n$, the closed-loop system has the properties

(i) the unique solutions $(x(\cdot), s(\cdot)) : [0, \infty) \to \mathbb{R}^{n+1}$, resp.
$(x(\cdot), s_1(\cdot), s_2(\cdot)) : [0, \infty) \to \mathbb{R}^{n+2}$, exist,

(ii) $\lim_{t \to \infty} s(t) = s_\infty$ resp. $\lim_{t \to \infty} s_i(t) = s_\infty$, $i = 1, 2$, exist and are finite,

(iii) $x(t)$ decays exponentially as t tends to ∞,

(iv) If (γ) is applied, then there exist $i \in \underline{N}$ and $t^* > 0$, so that $K(t) = K_i$ for all $t \geq t^*$.

Proof: (a):   Similar to part (a) of the proof of Theorem 4.2.6 it follows that for every $x_0 \in \mathbb{R}^n$ the closed-loop system (4.40) in feedback with one of (α)-(β) is well defined, and as in part (a) of the proof of Theorem 4.5.1 we obtain the existence of a unique solution $(x(\cdot), s(\cdot)) : [0, \omega) \to \mathbb{R}^{n+1}$, resp. $(x(\cdot), s_1(\cdot), s_2(\cdot)) : [0, \omega) \to \mathbb{R}^{n+2}$, maximally extended over $[0, \omega)$ for some $\omega \in (0, \infty]$ and $\lim_{t \to \infty} t_i = \omega$.

(b): Consider the adaptive strategy (γ) and use the notation

$$\bar{y}_1(t) := Zy(t), \qquad \bar{y}_2(t) := Wy(t).$$

We shall prove $s_1(\cdot), s_2(\cdot) \in L_\infty(0, \omega)$. To this end, choose $K_i \in \{K_1, \ldots, K_N\}$ such that $\sigma(Z^T CBZK_i) \subset \mathbb{C}_+$ and let $R = R^T \in \mathbb{R}^{(m-r) \times (m-r)}$ be the positive-definite solution of

$$RZ^T CBZK_i + (Z^T CBZK_i)^T R = I_{m-r}. \tag{4.42}$$

Define $\Lambda = RZ^T CBZ$, and choose $\alpha > 0$ sufficiently large so that

$$-\Lambda K_j - (\Lambda K_j)^T \leq \alpha I_{m-r} \quad \text{for all} \quad j \in \underline{N}. \tag{4.43}$$

By Theorem 3.2.3, for $S = \text{diag}\{\Lambda, I_r\}$, the system (4.40) is equivalent to the almost strictly positive real system (3.22) with

$$\bar{u}(t) = S[Z, W]^T u(t) = -S \begin{bmatrix} k_1(t)K(t), & 0 \\ 0 & k_2(t)I_r \end{bmatrix} \bar{y}(t).$$

Therefore we can apply Lemma 3.1.6 which yields, for all $t \in [0, \omega)$,

$$\frac{d}{dt} V(x(t)) \leq -\left\langle \bar{y}(t), \begin{bmatrix} k_1(t)\left[\Lambda K(t) + (\Lambda K(t))^T\right], & 0 \\ 0 & k_2(t)I_r \end{bmatrix} \bar{y}(t) \right\rangle$$

$$+ 2\|K\| \, \|\bar{y}(t)\|^2$$

$$= -k_1(t)\left\langle \bar{y}_1(t)^T \left[\Lambda K(t) + (\Lambda K(t))^T\right] \bar{y}_1(t)^T \right\rangle$$

$$+ 2\|K\|\|\bar{y}_1(t)\|^2 - [k_2(t) - 2\|K\|] \, \|\bar{y}_2(t)\|^2$$

for positive-definite $V(x) = \langle x, Px \rangle$ and some $K \in \mathbb{R}^{m \times m}$ as defined in Lemma 3.1.6. Integration over $[0, t_{lN}]$ yields

$$V(x(t_{lN})) \leq V(x(0)) - \int_0^{t_{lN}} k_1(\tau)\bar{y}_1(\tau)^T \left[\Lambda K(\tau) + (\Lambda K(\tau))^T\right] \bar{y}_1(\tau) d\tau$$

$$+ 2\|K\| \int_0^{t_{lN}} \dot{s}_1(\tau)d\tau - \int_0^{t_{lN}} [k_2(\tau) - 2\|K\|] \dot{s}_2(\tau)d\tau. \tag{4.44}$$

Suppose, without loss of generality, $i = N$ (otherwise integrate from a different time onwards) and we obtain for all $j \in \mathbb{N}$

$$-\int_{t_{jN}}^{t_{(j+1)N}} k_1(\tau)\bar{y}_1(\tau)^T \left[\Lambda K(\tau) + (\Lambda K(\tau))^T\right] \bar{y}_1(\tau)d\tau$$

$$\leq \alpha \int_{t_{jN}}^{t_{(j+1)N-1}} k_1(\tau)\dot{s}_1(\tau)d\tau - \int_{t_{(j+1)N-1}}^{t_{(j+1)N}} k_1(\tau)\dot{s}_1(\tau)d\tau$$

$$\leq \alpha T_{(j+1)N-1}[T_{(j+1)N-1} - T_{jN}]$$

$$-T_{(j+1)N}[T_{(j+1)N} - T_{(j+1)N-1}]$$

$$= T_{(j+1)N}^2 \left[\alpha\frac{T_{(j+1)N-1}^2 - T_{(j+1)N-1}T_{jN}}{T_{(j+1)N}^2} + \frac{T_{(j+1)N-1}}{T_{(j+1)N}} - 1\right]. \quad (4.45)$$

′ assumption on the growth rate of $\{T_i\}_{i\in\mathbb{N}}$, the right hand side in (4.45) tends to $-T_{(j+1)N}^2$ for $j \to \infty$. Thus there exists a $M_1 > 0$ such that

$$-\sum_{j=0}^{l-2} \int_{t_{jN}}^{t_{(j+1)N}} k_1(\tau)\bar{y}_1(\tau)^T \left[\Lambda K(\tau) + (\Lambda K(\tau))^T\right] \bar{y}_1(\tau)d\tau \leq M_1 \text{ for all } l \in \mathbb{N}.$$

(4.45) also yields

$$\frac{-1}{T_{lN} - T_0} \int_{t_{(l-1)N}}^{t_{lN}} k_1(\tau)\bar{y}_1(\tau)^T \left[\Lambda K(\tau) + (\Lambda K(\tau))^T\right] \bar{y}_1(\tau)d\tau$$

$$\leq \frac{T_{lN}^2}{T_{lN} - T_0} \left[\alpha\frac{T_{lN-1}^2 - T_{lN-1}T_{jN}}{T_{lN}^2} + \frac{T_{lN-1}}{T_{lN}} - 1\right]. \quad (4.46)$$

Inserting (4.45), (4.46) into (4.44) gives

$$V(x(t_{lN})) \leq V(x(0)) - \sum_{j=0}^{l-2} \int_{t_{jN}}^{t_{(j+1)N}} k_1(\tau)\bar{y}_1(\tau)^T \left[\Lambda K(\tau) + (\Lambda K(\tau))^T\right] \bar{y}_1(\tau)d\tau$$

$$- \int_{t_{(l-1)N}}^{t_{lN}} k_1(\tau)\bar{y}_1(\tau)^T \left[\Lambda K(\tau) + (\Lambda K(\tau))^T\right] \bar{y}_1(\tau)d\tau$$

$$+2\|K\|(T_{lN} - T_0) + 2\|K\|(T_{lN} - T_0) - \sum_{j=0}^{lN} T_j[T_j - T_{j-1}]$$

$$\leq V(x(0)) + M_1 + T_{lN}^2 \left[\alpha\frac{T_{lN-1}^2 - T_{lN-1}T_{jN}}{T_{lN}^2} + \frac{T_{lN-1}}{T_{lN}} - 1\right]$$

$$+4\|K\|[T_{lN} - T_0] - [T_{lN} - T_0]\sum_{j=0}^{lN} T_j\frac{T_j - T_{j-1}}{T_{lN} - T_0}$$

$$= V(x(0)) + M_1 + T_{lN}^2 \left[\alpha\frac{T_{lN-1}^2 - T_{lN-1}T_{jN}}{T_{lN}^2} + \frac{T_{lN-1}}{T_{lN}} - 1\right]$$

$$+[T_{lN} - T_0]\left[4\|K\| - \frac{T_{lN}^2 - T_{lN}T_{lN-1}}{T_{lN} - T_0} - \sum_{j=0}^{lN} T_j\frac{T_j - T_{j-1}}{T_{lN} - T_0}\right].$$

Since, by the growth requirement on $\{T_i\}_{i\in\mathbb{N}}$, the right hand side takes negative values as $l \to \infty$, thus contradicting the positiveness of the left hand side. This proves $s_1(\cdot), s_2(\cdot) \in L_\infty(0,\omega)$.

(c): Similar arguments as in (b) are used to prove boundedness of $s(\cdot)$ resp. $s_1(\cdot), s_2(\cdot)$, if the switching strategy (α) or (β) is applied. The proof is omitted.

(d): The remainder of the proof follows from Lemma 2.1.8 and similar arguments as in part (c) of the proof of Theorem 4.5.1, it is omitted.
This completes the proof of the theorem.   □

Remark 4.5.6
Using the ideas in the proof of Theorem 4.5.4, it is straightforward to state and prove an analogy to Theorem 4.2.4, where almost strictly positive real systems are considered, and Corollary 4.2.5 for piecewise constant feedback mechanisms.

4.6 Notes and References

Section 4.1: Morse (1983) conjectured that there does not exist a universal adaptive stabilizer of order 1 of the form $u(t) = f(k(t), y(t))$, $\dot{k}(t) = g(k(t), y(t))$, f and g differentiable, for the class of systems (4.1) with $cb \neq 0$. This would imply that the knowledge of the high-frequency gain is a necessary condition for universal adaptive stabilization. Nussbaum (1983) proved that the conjecture is true if f and g are required to be polynomials or rational functions. More importantly, Nussbaum introduced the following rich ·class of analytic controllers which are universal adaptive stabilizers for (4.1)

$$u(t) = [k(t)^2 + 1] \cdot N(k(t)) \cdot y(t), \quad \dot{k}(t) = y(t) \cdot [k(t)^2 + 1], \quad k(0) \in \mathbb{R}$$

$N(k)$ satisfies (4.3). Nussbaum's original example, the function $N_6(k)$ in Example 4.1.2, has been simplified by Willems and Byrnes (1984) and Morse (1984). The concept of scaling-invariant Nussbaum functions is due to Logemann and Owens (1988). The Nussbaum function (4.4) is due to Ilchmann and Townley (1992), and is of particular importance for relative degree two systems or nonlinearly perturbed systems. The proof of Lemma 4.1.4 is due to Mårtensson (1986,1987). The importance of the growth condition (4.5) has been discovered by Ryan (1991a) and Ilchmann and Logemann (1992). The sequence $\tau_{i+1} = \tau_i + e^{i^2}$, cf. Remark 4.1.6 (iv), was suggested by Byrnes and Willems (1984). The first part of the proof of Lemma 4.1.7 is from Ilchmann and Logemann (1992), the lemma is crucial for proving universal adaptive stabilization in case of $det(CB) \neq 0$. The concept of switching decision function has been introduced by Ilchmann and Owens (1991). Lemma 4.1.9 is implicitly contained in Ilchmann and Owens (1991b).

Section 4.2: Theorem 4.2.1 has a long history. With respect to the feedback laws (α)-(β) it has been proved by

(α) : Willems and Byrnes (1984) for single-input, single-output systems and $p = 2$; Mareels (1984) for single-input, single-output systems, $p \geq 1$, but a known bound on the high-frequency gain is required; Ilchmann et al. (1987) for multi-input, multi-output systems and $p = 2$.

(β) : Willems and Byrnes (1984) for single-input, single-output systems and $p = 2$; Ilchmann and Owens (1992) for multi-input, multi-output systems and $p \geq 1$.

(γ) : Byrnes and Willems (1984) and Mårtensson (1986) for $p = 2$, (both proofs are incomplete); Ilchmann and Logemann (1992) for $p \geq 1$.

Theorem 4.2.2 was conjectured by Morse (1986) and has been proved by Ilchmann and Townley (1992). The difficult part is that the systems class contains relative degree 1 and 2 systems. Under the additional assumption, that for every system belonging to (4.11) the sign of the high-frequency gain is positive,

Morse (1986) and Corless (1988, 1991) have proved that the feedback strategy $\dot{k}(t) = y(t)^2, u(t) = -k(t)y(t)$ is a universal adaptive stabilizer. Recently, Corless and Ryan (1992) have proved that the modification

$$u(t) = N(k(t))y(t), \qquad k(t) = \varepsilon y(t)^2 + \int\limits_0^t y(s)^2 ds + k(0), \quad k(0) \in \mathbb{R},$$

with $N(\cdot)$ a Nussbaum function, is a universal adaptive stabilizer for all relative degree 2 systems belonging to (4.11) provided that, additionally, some

$$\varepsilon > -\frac{1}{2}\frac{cAb}{cA^2b}$$

is *known* to the designer. It is not known whether this feedback strategy is actually a universal adaptive stabilizer for relative degree 1 systems belonging to (4.11).

The generalization mentioned in Remark 4.2.8 has firstly been considered in Ilchmann et al. (1987) and Owens et al. (1987).

Section 4.3: The idea of switching via a switching decision function has been introduced by Ilchmann and Owens (1991) for single-input, single-output, strictly proper, minimum phase systems. Theorem 4.3.1 has been proved in Ilchmann and Owens (1991a), the proof presented here is different and simpler. The definition of of the decision function $\psi(\cdot)$ has been slightly improved in the present text.

Section 4.4: The example in the introduction of Section 4.4 is from Hicks and Townley (1992), where they give explicit solutions for scalar systems and the switching strategy as in the example. The definition of λ-minimum phase systems is from Owens et al. (1987). The constant weighting factor has been introduced by in Owens et al. (1987) for the 'maximum controller' in Theorem 4.4.4, and in Logemann (1990) for $\dot{k}(t) = e^{\lambda t} y(t)^2$. The approach to find λ adaptively is due to Ilchmann and Owens (1990), and has been used for multivariable systems in the known sign case, for single-input, single-output sytems in the unknown sign case. Lemma 4.4.5 and Theorem 4.4.6 (for the modification of Theorem 4.2.1) is from Ilchmann and Logemann (1992).

Section 4.5: Piecewise constant gain adaptation has been introduced in Ilchmann and Owens (1991c). The proof of Theorem 4.5.1 for strictly proper, single-input single-output, systems (A, b, c) with $cb > 0$ and $p = 2$ is from Ilchmann (1992).

Exponential stabilization has also been achieved by a different piecewise constant feedback gain strategy for non-minimum phase systems by Miller and Davison (1989).

Many results of Section 4.2 can be extended to **infinite-dimensional** systems. Dahleh and Hopkins (1986), Kobayashi (1987), Byrnes (1987), and Dahleh (1988) have proved that $\dot{k}(t) = y(t)^2, u(t) = -N(k(t))y(t)$, where $N(\cdot)$ is a Nussbaum function, is a universal adaptive stabilizer for rather restricted classes. Theses results have been considerably extended by Logemann and Owens (1988), where an input-output theory is developed, by Logemann (1990), Logemann and Zwart (1991), and by Logemann and Mårtensson (1991). In Logemann and Ilchmann (1992), Theorem 4.2.1 with feedback (γ) has been proved via an input-output approach for a rather general class of systems. Exponential decay of the state $x(t)$, by using an exponential weighting factor $\dot{k}(t) = e^{\lambda t}y(t)^2$ (cf. Section 4.4), has been proved by Logemann (1990) for retarded systems which are λ-minimum phase.

High-gain stabilizable systems (by linear output feedback) are necessarily of relative degree smaller or equal to 2, see Theorem 2.2.6. It is possible to stabilize relative degree $p \geq 2$ systems by using multiparameter, dynamic, or derivative feedback controllers. Results on minimum phase systems which are in the spirit of this chapter have been achieved by: Mareels (1984), Mudgett and Morse (1989) for relative degree $p \geq 1$; Byrnes and Isidori (1986), Ilchmann (1991) for relative degree 2; Morse (1987) for relative degree 2; and Mårtensson (1986) and Miller and Davison (1991) for relative degree $p \geq 1$ using a different piecewise constant feedback strategy.

A theory of adaptive stabilization by discontinuous output feedback, and thus based on set valued maps and differential inclusions, has been developed by Ryan (1988, 1990, 1991).

Chapter 5

Universal Adaptive Tracking

In this chapter, two different adaptive tracking controllers are introduced. In Section 5.1, we show how an internal model, reduplicating the dynamics of the signals to be tracked, is connected in series with the unknown plant, so that essentially all universal adaptive stabilizers of Chapter 4 can be used to design a universal adaptive tracking controller, see Definition 1.1.2. The signals to be tracked belong to a class generated by a known differential equation.

In Section 5.2, we address the problem of λ-tracking, see Definition 1.1.2. By sacrificing asymptotic tracking for the weaker (but nonetheless practical) requirement of tracking with error asymptotic to a ball of arbitrarily small prespecified radius $\lambda > 0$, we design a controller which has, compared to the tracking controllers of Section 5.1, the following advantages:

- no internal model is invoked,
- simple linear feedback $u(t) = -N(k(t))[y(t) - y_{\text{ref}}(t)]$ as in Chapter 4 is used,
- the gain adaptation is changed by introducing a dead-zone depending on $\lambda > 0$, however the simplicity of the adaptation law is preserved,
- the class of reference signals is the large class $\mathcal{W}^{1,\infty}(\mathbb{R}, \mathbb{R}^m)$ of bounded functions that are absolutely continuous on compact intervals of $\mathbb{R}$ and have essentially bounded derivatives,
- the controller tolerates large classes of nonlinearities in the state and input of the nominal system, see Section 6.3,
- the controller can cope with output corrupted noise belonging to the same class as the reference signals, i.e. $\mathcal{W}^{1,\infty}(\mathbb{R}, \mathbb{R}^m)$, see also Section 6.3.

5.1   Asymptotic tracking

In this section, we consider the asymptotic tracking problem for various classes of multivariable, linear, minimum phase systems of the form

$$\left.\begin{array}{rcl} \dot{x}(t) & = & Ax(t) + Bu(t) \quad , x(0) \in \mathbb{R}^n \\ y(t) & = & Cx(t) + Du(t), \end{array}\right\} \qquad (5.1)$$

with $(A, B, C, D) \in \mathbb{R}^{n \times n} \times \mathbb{R}^{n \times m} \times \mathbb{R}^{m \times n} \times \mathbb{R}^{m \times m}$, and the class of reference signals

$$\mathcal{Y}_{\text{ref}} := \left\{ y_{\text{ref}}(\cdot) \in \mathcal{C}^\infty(\mathbb{R}, \mathbb{R}^m) \mid \alpha(\tfrac{d}{dt}) y_{\text{ref}}(t) \equiv 0 \right\},$$

where $\alpha(\cdot) \in \mathbb{R}[s]$. In other words, the reference class consists of all signals satisfying a scalar differential equation with constant coefficients, therefore $y_{\text{ref}}(t)$ may consist of signals, and its linear combinations, of the form

$$a_0, \ \sin(\omega t), \ \cos(\varphi t), \ \left(\frac{1}{(k-1)!} t^{k-1} v_1 + \ldots + t v_{k-1} + v_k \right) e^{\lambda t}.$$

To achieve that the output tracks an arbitrary reference signal, the main idea is to connect the unknown system $G(\cdot) \in \mathbb{R}(s)^{m \times m}$, belonging to a certain class Σ, in series with an internal model $\frac{\beta(\cdot)}{\alpha(\cdot)} I_m$ reduplicating the dynamics of the reference class, and applying one of the universal adaptive stabilizers of Chapter 4 to the composed system $\bar{G}(s) = \frac{\beta(s)}{\alpha(s)} \cdot G(s)$. See Figure 5.1. The *Internal Model Principle* is described as follows: "A regulator is structurally stable only if the controller utilizes feedback of the regulated variable, and incorporates in the feedback loop a suitable reduplicated model of the dynamic structure of the exogenous signals which the regulator is required to process." Wonham (1979).

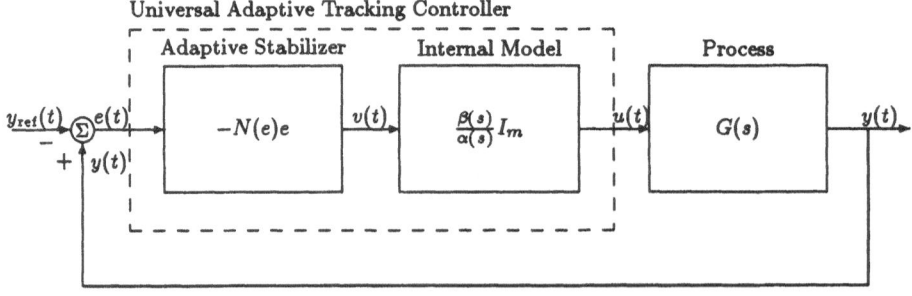

Figure 5.1: Asymptotic tracking controller with internal model

We will see that if $\beta(\cdot)$ is chosen properly ($\beta(\cdot)$ Hurwitz with $\deg \beta(\cdot) = \deg \alpha(\cdot)$ and, possibly, $\lim_{s\to\infty} \frac{\beta(s)}{\alpha(s)} > 0$), then the relevant structural properties of $G(\cdot)$ are inherited by $\bar{G}(\cdot)$ and the universal adaptive tracking problem can be transformed into a universal adaptive stabilizing problem.

To this end, we first derive a state space representation for the series connection

$$\bar{G}(s) = \frac{\beta(s)}{\alpha(s)} I_m \cdot G(s)$$

where $G(\cdot)$ denotes the transfer function of (5.1), and $\beta(\cdot), \alpha(\cdot) \in \mathbb{R}[s]$ with $\deg \beta(\cdot) = \deg \alpha(\cdot)$. Let $(\hat{A}, \hat{b}, \hat{c}, \hat{d}) \in \mathbb{R}^{p \times p} \times \mathbb{R}^p \times \mathbb{R}^{1 \times p} \times \mathbb{R}$ be a minimal state space realization of $\frac{\beta(s)}{\alpha(s)}$ and

$$\hat{d} \quad := \quad \lim_{s\to\infty} \frac{\beta(s)}{\alpha(s)}.$$

Thus, a minimal state space realization of $\frac{\beta(s)}{\alpha(s)} I_m$, i.e. the *internal model*, is given by

$$\left.\begin{array}{rcl} \dot{\xi}(t) &=& \hat{A}^*\xi(t) + \hat{B}^* v(t), \quad \xi(0) = \xi_0 \\ u(t) &=& \hat{C}^*\xi(t) + \hat{d}I_m v(t) \end{array}\right\} \tag{5.2}$$

with

$$\begin{array}{rcl} \hat{A}^* &=& \text{diag}\{\hat{A}, \ldots, \hat{A}\} \in \mathbb{R}^{mp \times mp}, \\ \hat{B}^* &=& \text{diag}\{\hat{b}, \ldots, \hat{b}\} \in \mathbb{R}^{mp \times m}, \\ \hat{C}^* &=& \text{diag}\{\hat{c}, \ldots, \hat{c}\} \in \mathbb{R}^{m \times mp}. \end{array}$$

A straightforward calculation yields that a stabilizable and detectable state space realization of $\bar{G}(s) = \frac{\beta(s)}{\alpha(s)} I_m \cdot G(s)$ is

$$\left.\begin{array}{rcl} \dot{\bar{x}}(t) &=& \bar{A}\bar{x}(t) + \bar{B}v(t) \quad, \bar{x}(0) = (x_0^T, \xi_0^T)^T \\ y(t) &=& \bar{C}\bar{x}(t) + \bar{D}v(t), \end{array}\right\} \tag{5.3}$$

with

$$\bar{A} = \begin{bmatrix} A & B\hat{C}^* \\ 0 & \hat{A}^* \end{bmatrix}, \quad \bar{B} = \begin{bmatrix} \hat{d}B \\ \hat{B}^* \end{bmatrix}, \quad \bar{C} = \begin{bmatrix} C, D\hat{C}^* \end{bmatrix}, \quad \bar{D} = \hat{d}D, \quad \bar{x} = \begin{bmatrix} x \\ \xi \end{bmatrix}.$$

Using this notation, it is easy to see that the following properties of $G(\cdot)$ are inherited by $\bar{G}(\cdot)$.

Lemma 5.1.1

(i) $\bar{G}(\cdot)$ is minimum phase if $G(\cdot)$ is minimum phase and $\beta(\cdot)$ is Hurwitz,

(ii) $\bar{D} + \bar{D}^T = \hat{d}(D + D^T)$,

(iii) If $D = 0$, then

$$
\begin{aligned}
\bar{C}\bar{B} &= \hat{d}CB, \\
\bar{C}\bar{A}\bar{B} &= \hat{d}CAB + CB\hat{C}^*\hat{B}^*, \\
\bar{C}\bar{A}^2\bar{B} &= \hat{d}CA^2B + CB\hat{C}^*\hat{A}^*\hat{B}^*.
\end{aligned}
$$

It is also obvious that $\mathrm{rk}\,\bar{B} = m \leq n$ if $\mathrm{rk}\,B = m \leq n$. However, strict positive realness of $G(\cdot)$ does not imply strict positive realness of $\bar{G}(\cdot)$ since $\alpha(\cdot)$ may have poles on $i\mathbb{R}$.

Before proving the main result, we require a technical lemma. To this end, $\alpha(\cdot) \in \mathbb{R}[s]$ is factorized into $\alpha(\cdot) = \alpha^-(\cdot)\alpha^+(\cdot)$ with

$$
\begin{aligned}
\alpha^-(s_0) = 0 &\iff \mathrm{Re}\,s_0 < 0 \\
\alpha^+(s_0) = 0 &\iff \mathrm{Re}\,s_0 \geq 0
\end{aligned}
$$

and, according to this factorization, we define

$$
\begin{aligned}
\mathcal{Y}_{\mathrm{ref}}^- &:= \left\{ y_{\mathrm{ref}}(\cdot) \in \mathcal{C}^\infty(\mathbb{R}, \mathbb{R}^m) \mid \alpha^-(\tfrac{d}{dt})y_{\mathrm{ref}}(t) \equiv 0 \right\}, \\
\mathcal{Y}_{\mathrm{ref}}^+ &:= \left\{ y_{\mathrm{ref}}(\cdot) \in \mathcal{C}^\infty(\mathbb{R}, \mathbb{R}^m) \mid \alpha^+(\tfrac{d}{dt})y_{\mathrm{ref}}(t) \equiv 0 \right\}.
\end{aligned}
$$

Clearly, $\mathcal{Y}_{\mathrm{ref}} = \mathcal{Y}_{\mathrm{ref}}^- \oplus \mathcal{Y}_{\mathrm{ref}}^+$.

Lemma 5.1.2
If the system (5.1) is minimum phase and $\bar{A}$, $\bar{C}$ are given as in (5.3), then for every $y_{\mathrm{ref}}^+(\cdot) \in \mathcal{Y}_{\mathrm{ref}}^+$ there exists a $\tilde{x}_0 \in \mathbb{R}^{n+mp}$ and $M > 0$ such that

$$
\left.
\begin{aligned}
\dot{\tilde{x}}(t) &= \bar{A}\tilde{x}(t), \quad \tilde{x}(0) = \tilde{x}_0 \\
y_{\mathrm{ref}}(t) &= \bar{C}\tilde{x}(t)
\end{aligned}
\right\}
\tag{5.4}
$$

and

$$
\|\tilde{x}(t)\| \leq M \left[1 + \max_{s \in [0,t]} \|y_{\mathrm{ref}}^+(s)\| \right] \quad \text{for all} \quad t \geq 0.
\tag{5.5}
$$

Proof: (a): From the classical theory of ordinary linear differential equations it is known that every (real) $y_{\mathrm{ref}}^+(\cdot)$ is a (possibly complex) linear combination of

$$
e^{\lambda_1 t}, te^{\lambda_1 t}, \ldots, t^{m_1-1}e^{\lambda_1 t}, \ldots, e^{\lambda_\nu t}, te^{\lambda_\nu t}, \ldots, t^{m_\nu-1}e^{\lambda_\nu t},
$$

where $\lambda_1, \ldots, \lambda_\nu$ are the zeros of $\alpha^+(\cdot)$ with algebraic multiplicity $m_1, \ldots, m_\nu$, respectively.

(b): First we shall determine $\tilde{x}_0 \in \mathbb{C}^{n+mp}$ to solve (5.4) for the case

$$
y_{\mathrm{ref}}^+(t) = t^k e^{\lambda t}\psi
$$

where $\lambda \in \{\lambda_1, \ldots, \lambda_\nu\}$, k smaller than the algebraic multiplicity of λ, and $\psi \in \mathbb{C}^m$.

Partioning $\tilde{x} = (z^T, \zeta^T)^T$, (5.4) is equivalent to

$$\left.\begin{array}{rcl}
\dot{z}(t) & = & Az(t) + B\hat{C}^*\zeta(t), \quad z(0) = z_0 \\
\dot{\zeta}(t) & = & \hat{A}^*\zeta(t), \quad\quad\quad\quad \zeta(0) = \zeta_0 \\
y_{\text{ref}}^+(t) & = & Cz(t) + D\hat{C}^*\zeta(t).
\end{array}\right\} \tag{5.6}$$

We shall show that

$$\hat{C}^*\zeta(t) = \sum_{i=0}^{k} \hat{\gamma}_i t^i e^{\lambda t} \quad \text{and} \quad z(t) = \sum_{i=0}^{k} \gamma_i t^i e^{\lambda t}$$

satisfy the first and third equation in (5.6), where the coefficients $\gamma_i, \hat{\gamma}_i$ are defined as follows

$$\begin{bmatrix} \gamma_k \\ \hat{\gamma}_k \end{bmatrix} := \begin{bmatrix} A - \lambda I_n & B \\ C & D \end{bmatrix}^{-1} \begin{bmatrix} 0 \\ \psi \end{bmatrix}$$

$$\begin{bmatrix} \gamma_i \\ \hat{\gamma}_i \end{bmatrix} := \begin{bmatrix} A - \lambda I_n & B \\ C & D \end{bmatrix}^{-1} \begin{bmatrix} (i+1)\gamma_{i+1} \\ 0 \end{bmatrix} \quad \text{for} \quad i = k-1, \ldots, 0.$$

Since $\lambda \in \overline{\mathbb{C}}_+$ and (A, B, C, D) is minimum phase, Proposition 2.1.2 yields that the inverse above is well defined. Inserting the ansatz of $\hat{C}^*\zeta(t)$ and $z(t)$ into (5.6) yields that (5.6) is equivalent to

$$\lambda \sum_{i=0}^{k} \gamma_i t^i + \sum_{i=0}^{k-1} (i+1)\gamma_{i+1} t^i = A \sum_{i=0}^{k} \gamma_i t^i + B \sum_{i=0}^{k} \hat{\gamma}_i t^i,$$

$$t^k \psi = C \sum_{i=0}^{k} \gamma_i t^i + D \sum_{i=0}^{k} \hat{\gamma}_i t^i,$$

which is, by comparing coefficients, equivalent to

$$\begin{bmatrix} A - \lambda I_n & B \\ C & D \end{bmatrix} \begin{bmatrix} \gamma_k, & \cdots & , \gamma_0 \\ \hat{\gamma}_k, & \cdots & , \hat{\gamma}_0 \end{bmatrix} = \begin{bmatrix} 0, & k\gamma_k, & \cdots & , \gamma_1 \\ \psi, & 0, & \cdots & , 0 \end{bmatrix}.$$

This is just the definition of γ_i and $\hat{\gamma}_i$, $i = 0, \ldots, k$, and therefore, it remains to show the existence of a solution $\zeta(t)$ of

$$\dot{\zeta}(t) = \hat{A}^*\zeta(t), \quad \zeta(0) = \zeta_0$$

$$\hat{C}^*\zeta(t) = \sum_{i=0}^{k} \hat{\gamma}_i t^i e^{\lambda t}.$$

Since $\sum_{i=0}^{k} \hat{\gamma}_i t^i e^{\lambda t} \in \left\{ e^{\hat{A}^* t} \zeta \,\middle|\, \zeta \in \mathbb{C}^{mp} \right\}$, and since $(\hat{A}^*, \hat{C}^*)$ is observable, there exists a $\zeta_0 \in \mathbb{C}^{mp}$ so that the equation above holds true.

So far a solution $z(t), \zeta(t)$, $\tilde{x}_0 = (x_0^T, \xi_0^T)^T$ of (5.4) has been constructed which might have complex values.

(c):   To obtain a real $\tilde{x}_0$ for arbitrary $y_{\text{ref}}^+(\cdot) \in \mathcal{Y}_{\text{ref}}^+$, note that the real $y_{\text{ref}}^+(\cdot)$ can be written as a sum of (possibly complex) terms investigated in (b). The existence of $\tilde{x}_0$ to solve (5.4) then follows by superposition and taking, if necessary, the real part of $\tilde{x}_0$ as constructed in (b).

(d):   It remains to prove (5.5). By Lemma 5.1.1, $(\bar{A}, \bar{B}, \bar{C}, \bar{D})$ is minimum phase and therefore, in view of Proposition 2.1.2, $(\bar{A}, \bar{C})$ is detectable. Thus there exists an $\bar{H} \in \mathbb{R}^{(n+mp) \times m}$ such that $\sigma(\bar{A} - \bar{H}\bar{C}) \subset \mathbb{C}_-$. Rewriting $\dot{\tilde{x}}(t) = \bar{A}\tilde{x}(t)$ as

$$\dot{\tilde{x}}(t) = [\bar{A} - \bar{H}\bar{C}]\tilde{x}(t) + \bar{H}y_{\text{ref}}^+(t),$$

it follows from the stability of $\bar{A} - \bar{H}\bar{C}$ that there exist $\hat{M}, \omega > 0$ such that

$$\|\tilde{x}(t)\| \quad \leq \quad \hat{M}e^{-\omega t}\|\tilde{x}(0)\| + \hat{M} \int_0^t e^{-\omega(t-s)}\|\bar{H}\|\|y_{\text{ref}}^+(s)\|ds$$

$$\leq \quad \hat{M}e^{-\omega t}\|\tilde{x}(0)\| + \frac{\hat{M}\|\bar{H}\|}{\omega} \max_{s \in [0,t]} \|y_{\text{ref}}^+(s)\|$$

for all $t \geq 0$. This prove (5.5) for $M := \hat{M}\|\tilde{x}(0)\| + \frac{\hat{M}\|\bar{H}\|}{\omega}$, and completes the proof of the
lemma. $\square$

We are now in a position to present the main result of this section, that is, all universal adaptive stabilizers introduced in Chapter 4 in series with an appropriate internal model, reduplicating the dynamics of the reference signals, can be used to design a universal adaptive tracking controller.

Theorem 5.1.3
Consider a minimum phase system

$$\left. \begin{array}{rcl} \dot{x}(t) & = & Ax(t) + Bu(t) \quad , x(0) \in \mathbb{R}^n \\ y(t) & = & Cx(t) + Du(t), \end{array} \right\} \tag{5.7}$$

with $(A, B, C, D) \in \mathbb{R}^{n \times n} \times \mathbb{R}^{n \times m} \times \mathbb{R}^{m \times n} \times \mathbb{R}^{m \times m}$, and the class of reference signals

$$\mathcal{Y}_{\text{ref}} := \left\{ y_{\text{ref}}(\cdot) \in C^\infty(\mathbb{R}, \mathbb{R}^m) \mid \alpha(\tfrac{d}{dt})y_{\text{ref}}(t) \equiv 0 \right\},$$

for some $\alpha(\cdot) \in \mathbb{R}[s]$.
Choose a Hurwitz polynomial $\beta(\cdot) \in \mathbb{R}[s]$ with $\deg \beta(\cdot) = \deg \alpha(\cdot)$, and a minimal realization of $\frac{\beta(\cdot)}{\alpha(\cdot)}I_m$ as in (5.2).
Suppose (5.7) belongs to a class Σ considered in Theorem 4.2.1, 4.2.2, 4.2.6,

4.3.1, 4.3.3, 4.4.3, 4.4.6, 4.5.1, 4.5.3, or 4.5.5, with corresponding feedback strategy

$$
\begin{aligned}
u(t) &= f_\Sigma\left(k(\cdot)|_{[0,t]}, y(\cdot)|_{[0,t]}\right)(t) \cdot y(t) \\
k(t) &= g_\Sigma\left(t, y(\cdot)|_{[0,t]}\right).
\end{aligned}
$$

If Σ requires $\sigma(CB) \subset \mathbb{C}_+$ or $D + D^T > 0$, then we assume, in addition, $\lim_{s\to\infty} \frac{\beta(s)}{\alpha(s)} = \hat{d} > 0$. If Σ coincides with (4.11), then the second Markov parameter of $\frac{\beta(s)}{\alpha(s)}$ is assumed to be non-positive. (This can always be achieved by rescaling $\beta(\cdot)$ with -1.)

Under these assumptions, the internal model

$$
\left.
\begin{aligned}
\dot{\xi}(t) &= \hat{A}^*\xi(t) + \hat{B}^* v(t), \quad \xi(0) = \xi_0 \\
u(t) &= \hat{C}^*\xi(t) + \hat{d}I_m v(t)
\end{aligned}
\right\}
\tag{5.8}
$$

in series connection with

$$
\left.
\begin{aligned}
e(t) &= y(t) - y_{\text{ref}}(t) \\
v(t) &= f_\Sigma\left(k(\cdot)|_{[0,t]}, e(\cdot)|_{[0,t]}\right)(t)e(t) \\
k(t) &= g_\Sigma\left(t, e(\cdot)|_{[0,t]}\right)
\end{aligned}
\right\}
\tag{5.9}
$$

is a universal adaptive tracking controller for the class Σ. More precisely, for arbitrary $x_0 \in \mathbb{R}^n, \xi_0 \in \mathbb{R}^{mp}$, and $y_{\text{ref}}(\cdot) \in \mathcal{Y}_{\text{ref}}$, the closed-loop system (5.7)-(5.9) has the properties

(i) the unique solution $(x(\cdot), \xi(\cdot), k(\cdot)) : [0, \infty) \to \mathbb{R}^{n+mp+1}$ exists,

(ii) $\lim_{t\to\infty} k(t) = k_\infty$ exists and is finite,

(iii) $\|x(t)\| + \|\xi(t)\| \le M[1 + \|y_{\text{ref}}(t)\|]$ for all $t \ge 0$ and some $M > 0$,

(iv) $\lim_{t\to\infty} \|y(t) - y_{\text{ref}}(t)\| = 0$.

Depending on the system class Σ and the feedback strategy chosen, we also obtain additional results, e.g. exponential convergence of $y(t)$ to $y_{\text{ref}}(t)$ in case of Theorem 4.5.1, we do not list them explicitly.

Proof: (a): Define

$$
x_e(t) := \bar{x}(t) - \tilde{x}(t)
$$

where $\bar{x}(t), \tilde{x}(t)$ satisfy (5.3), (5.4), respectively, and decompose

$$
y_{\text{ref}}(t) = y_{\text{ref}}^-(t) + y_{\text{ref}}^+(t) \quad \text{for} \quad y_{\text{ref}}^-(\cdot) \in \mathcal{Y}_{\text{ref}}^- \quad \text{and} \quad y_{\text{ref}}^+(\cdot) \in \mathcal{Y}_{\text{ref}}^+.
$$

By Lemma 5.1.2, $x_e(t)$ is the solution of

$$
\left.
\begin{aligned}
\dot{x}_e(t) &= \bar{A}x_e(t) + \bar{B}v(t) &, x_e(0) = (x_0^T, \xi_0^T)^T - \tilde{x}_0 \\
e(t) &= \bar{C}x_e(t) + \bar{D}v(t) - y_{\text{ref}}^-(t).
\end{aligned}
\right\}
\tag{5.10}
$$

If (A, b, c) belongs to (4.11), then it follows from Theorem 2.2.6, Lemma 5.1.1 (iii), and the assumption $\hat{d} > 0$, that $(\bar{A}, \bar{B}, \bar{C})$ belongs to (4.11) as well. Every other assumption made in the theorem on (A, B, C, D) carries over, by Lemma 5.1.1, to $(\bar{A}, \bar{B}, \bar{C}, \bar{D})$. Thus, if $y_{\text{ref}}^-(\cdot) \equiv 0$, we can apply the stabilization strategy of Chapter 4 (with $y(t)$ replaced by $e(t)$) to (5.10). This proves $\lim_{t \to \infty} x_e(t) = 0$ and the assertions (i), (ii) follow.

If $y_{\text{ref}}^-(t) \not\equiv 0$, then $y_{\text{ref}}^-(t)$ can be viewed as an exponentially bounded disturbance of the error measurement and the proof of previous adaptive stabilizers goes through with straightforward modifications.

If $D + D^T > 0$, then $f_\Sigma \left(k(\cdot)|_{[0,t]}, e(\cdot)|_{[0,t]} \right)(t)$ in the feedback law satisfies

$$\lim_{t \to \infty} f_\Sigma \left(k(\cdot)|_{[0,t]}, e(\cdot)|_{[0,t]} \right)(t) > 0$$

and we have

$$e(t) = \left[I_m + f_\Sigma \left(k(\cdot)|_{[0,t]}, e(\cdot)|_{[0,t]} \right)(t)D \right]^{-1} \bar{C} x_e(t)$$

for t sufficiently large. Therefore, by $\lim_{t \to \infty} x_e(t) = 0$, (iv) follows.

(b):   Using Lemma 5.1.1 and the boundedness of $x_e(\cdot)$, we have

$$
\begin{aligned}
\tfrac{1}{2} \left(\|x(t)\| + \|\xi(t)\| \right) \leq \|\bar{x}(t)\| \quad &= \quad \|\tilde{x}(t) + x_e(t)\| \\
&\leq \quad M_1 + \|\tilde{x}(t)\| \\
&\leq \quad M_2[1 + \max_{s \in [0,t]} \|y_{\text{ref}}^+(s)\|] \\
&\leq \quad M_3[1 + \|y_{\text{ref}}^+(t)\|] \\
&\leq \quad M_4[1 + \|y_{\text{ref}}(t)\|].
\end{aligned}
$$

for some $M_1, \ldots, M_4 > 0$. This completes the proof.   □

5.2 λ-Tracking

In this section, we solve the λ-tracking problem for various classes of multivariable, linear, minimum phase systems of the form

$$
\left.
\begin{aligned}
\dot{x}(t) &= Ax(t) + Bu(t) \quad, x(0) \in \mathbb{R}^n \\
y(t) &= Cx(t)
\end{aligned}
\right\}
\tag{5.11}
$$

with $(A, B, C) \in \mathbb{R}^{n \times n} \times \mathbb{R}^{n \times m} \times \mathbb{R}^{m \times n}$ and the class of reference signals is

$$\mathcal{Y}_{\text{ref}} = \mathcal{W}^{1,\infty}(\mathbb{R}, \mathbb{R}^m).$$

The feedback law will be the same as for the Willems-Byrnes controller, i.e. linear feedback of the form

$$
\begin{aligned}
u(t) &= f(k(t))e(t), \\
e(t) &= y(t) - y_{\text{ref}}(t),
\end{aligned}
$$

for $f(k) = -k$ or $f(k) = -N(k)$, $N(\cdot)$ a Nussbaum function. Only the gain adaptation takes into account the parameter $\lambda > 0$ and introduces a dead-zone. In the single-input, single-output case we can, for example, choose

$$
\dot{k}(t) = d_\lambda(e(t))|e(t)|, \quad k(0) = k_0 \in \mathbb{R}
$$

where $d_\lambda(\cdot) : \mathbb{R}^m \to [0, \infty)$ denotes the 'distance' function

$$
d_\lambda(e) := \begin{cases} \|e\| - \lambda & , \text{ if } \|e\| \geq \lambda \\ 0 & , \text{ if } \|e\| < \lambda. \end{cases} \tag{5.12}
$$

Before proving the first result, a technical lemma is required. The lemma is proved for a more general version than needed in Theorem 5.2.2, since it will be used in Section 6.3 in a more general context.

Lemma 5.2.1
Let $A_3 \in \mathbb{R}^{(n-m) \times m}$, $A_4 \in \mathbb{R}^{(n-m) \times (n-m)}$ with $\sigma(A_4) \subset \mathbb{C}_-$, $\omega \in (t_0, \infty]$, $z_0 \in \mathbb{R}^{n-m}$. If

$$
\theta(\cdot) : [t_0, \omega) \to \mathbb{R}^m \quad \text{and} \quad h(\cdot) : [t_0, \omega) \to \mathbb{R}^{n-m}
$$

are locally integrable functions satisfying

$$
\|h(t)\| \leq \hat{h}[1 + \|\theta(t)\|] \quad \text{for almost all} \quad t \in [t_0, \omega)
$$

for some $\hat{h} \geq 0$, then there exists $M > 0$ such that the solution $z(\cdot) : [t_0, \omega) \to \mathbb{R}^{n-m}$ of

$$
\dot{z}(t) = A_4 z(t) + A_3 \theta(t) + h(t), \quad z(t_0) = z_0 \tag{5.13}
$$

satisfies

$$
\int_{t_0}^t \|\theta(s)\| \cdot \|z(s)\| ds \leq M \int_{t_0}^t \left[\|\theta(s)\| + |\theta(s)\|^2 \right] ds \quad \text{for all} \quad t \in [t_0, \omega).
$$

Proof: Since A_4 is exponentially stable, there exist $M_1, \mu > 0$ such that variation of constants applied to (5.13) yields

$$
\|z(t)\| \leq M_1 e^{-\mu(t-t_0)} \|z_0\| + M_1 \int_{t_0}^t e^{-\mu(t-s)} \left[\|A_3\| \|\theta(s)\| + \|h(s)\| \right] ds,
$$

and so, by using the assumption on $\|h(t)\|$,

$$
\begin{aligned}
\|z(t)\| &\leq M_1 e^{-\mu(t-t_0)}\|z_0\| + M_1 \hat{h}\mu^{-1} + M_1\left[\|A_3\| + \hat{h}\right]\mathcal{L}(\|\theta(\cdot)\|)(t) \\
&\leq M_2\left[1 + \mathcal{L}(\|\theta(\cdot)\|)(t)\right],
\end{aligned}
$$

where $M_2 := M_1\left[\|z_0\| + \hat{h}(1 + \mu^{-1}) + \|A_3\|\right]$ and $\mathcal{L}$ denotes the operator

$$
\mathcal{L} : \varphi(\cdot) \mapsto \left(t \mapsto \int_{t_0}^{t} e^{-\mu(t-s)}\varphi(s)ds\right).
$$

Therefore, using Hölder's inequality

$$
\begin{aligned}
\int_{t_0}^{t} \|z(s)\|\|\theta(s)\|ds &\leq M_2\int_{t_0}^{t}\|\theta(s)\|\left[1 + \mathcal{L}(\|\theta(\cdot)\|)(s)\right]ds \\
&\leq M_2\int_{t_0}^{t}\|\theta(s)\|ds + M_2\|\mathcal{L}(\|\theta(\cdot)\|)(\cdot)\|_{L_2(t_0,t)} \cdot \|\theta(\cdot)\|_{L_2(t_0,t)}.
\end{aligned}
$$

By Theorem 6.5.54 in Vidyasagar (1978), we have

$$
\|\mathcal{L}(\|\theta(\cdot)\|)(\cdot)\|_{L_2(t_0,t)} \leq \|\mathcal{L}\|_{L_2(t_0,t)}\|\theta(\cdot)\|_{L_2(t_0,t)} \leq \frac{1}{\mu}\|\theta(\cdot)\|_{L_2(t_0,t)}
$$

for all $t \in [t_0, \omega)$. This proves the result for $M := M_2[1 + \mu^{-1}]$. $\square$

We are now in a position to prove a main result, that is adaptive λ-tracking for the class of single-input, single-output, minimum phase systems with high-frequency gain $cb > 0$ or $cb \neq 0$. Finite escape times of the nonlinear closed-loop system do not exist, all states are bounded, and the error between output and reference signal tends to the interval $[-\lambda, +\lambda]$, as $t \to \infty$, where $\lambda > 0$ is prespecified and determines the dead-zone in the gain adaptation.

Theorem 5.2.2
Suppose

$$
\left.\begin{aligned}
\dot{x}(t) &= Ax(t) + bu(t), \quad x(0) = x_0 \\
y(t) &= cx(t),
\end{aligned}\right\} \tag{5.14}
$$

with $(A, b, c) \in \mathbb{R}^{n \times n} \times \mathbb{R}^n \times \mathbb{R}^{1 \times n}$, is minimum phase. Let $\lambda > 0$, $N(\cdot) : \mathbb{R} \to \mathbb{R}$ a Nussbaum function and $y_{\text{ref}}(\cdot) \in \mathcal{W}^{1,\infty}(\mathbb{R}, \mathbb{R}^m)$. If the adaptation law

$$
\dot{k}(t) = \begin{cases} (|e(t)| - \lambda)|e(t)| & , \text{if} \ \ |e(t)| \geq \lambda \\ 0 & , \text{if} \ \ |e(t)| < \lambda \end{cases} \quad, k(0) = k_0
$$

together with one of the feedback laws

$(\alpha):$ $u(t) = -k(t)e(t)$, if $cb > 0$

$(\beta):$ $u(t) = -N(k(t))e(t)$, if $cb \neq 0$

where

$$e(t) = y(t) - y_{\text{ref}}(t),$$

is applied to (5.14), for arbitrary $x_0 \in \mathbb{R}^n$, $k_0 \in \mathbb{R}$, then the closed-loop system has the properties

(i) there exists a unique solution $(x(\cdot), k(\cdot)) : [0, \infty) \to \mathbb{R}^{n+1}$,

(ii) $\lim_{t \to \infty} k(t) = k_\infty$ exists and is finite,

(iii) $x(\cdot), k(\cdot) \in L_\infty(0, \infty)$,

(iv) the error $e(t) = y(t) - y_{\text{ref}}(t)$ approaches the interval $[-\lambda, \lambda]$ as $t \to \infty$.

Proof: We only consider the feedback (β), the proof for (α) simply follows by replacing $N(k)$ by k in the following proof.

(a):   Since the right hand side of the closed-loop system

$$\begin{aligned}
\dot{x}(t) &= [A - N(k(t))bc]x(t) + N(k(t))by_{\text{ref}}(t) &, x(0) = x_0 \\
\dot{k}(t) &= d_\lambda(e(t))|e(t)| &, k(0) = k_0
\end{aligned}$$

where $d_\lambda(\cdot)$ is defined in (5.12), is piecewise right continuous and locally Lipschitz, it follows from the classical theory of ordinary differential equations, that for every $(x_0, k_0) \in \mathbb{R}^{n+1}$ there exists a unique solution $(x(\cdot), k(\cdot)) : [0, \omega) \to \mathbb{R}^{n+1}$, which can be maximally extended over $[0, \omega)$ for some $\omega > 0$.

(b):   By the state space transformation $S^{-1}x = (y^T, z^T)^T$ given in Lemma 2.1.3, the closed-loop system is equivalent to

$$\left.\begin{aligned}
\dot{e}(t) &= [A_1 - N(k(t))cb]e(t) + A_2 z(t) + g_1(t) &, e(0) = Cx_0 - y_{\text{ref}}(0) \\
\dot{z}(t) &= \quad\quad A_3 e(t) + A_4 z(t) + g_2(t) &, z(0) = Nx_0 \\
\dot{k}(t) &= \quad d_\lambda(e(t))|e(t)| &, k(0) = k_0.
\end{aligned}\right\} \quad (5.15)$$

where

$$\begin{aligned}
g_1(t) &:= -\dot{y}_{\text{ref}}(t) \\
g_2(t) &:= A_1 y_{\text{ref}}(t) - \dot{y}_{\text{ref}}(t).
\end{aligned}$$

(c):   We shall derive an inequality for $e(t)$. Differentiation of the following C^1-function (a Lyapunov-like candidate)

$$V_\lambda(\cdot) : \mathbb{R}^m \to [0, \infty), \quad e \mapsto \frac{1}{2}d_\lambda(e)^2 = \begin{cases} \frac{1}{2}(|e| - \lambda)^2 &, \text{if } |e| \geq \lambda \\ 0 &, \text{if } |e| < \lambda \end{cases}$$

along the solution component $e(t)$ of (5.15) yields

$$\frac{d}{dt}V_\lambda(e(t)) = \theta(t)\dot{e}(t) \quad \text{for all} \quad t \in [0,\omega),$$

where, for notational convenience, we have introduced the continuous function

$$\theta(\cdot) : \mathbb{R} \to \mathbb{R}, \quad t \mapsto \theta(t) := \begin{cases} \frac{|e(t)|-\lambda}{|e(t)|}e(t) & \text{, if} \quad |e(t)| \geq \lambda \\ 0 & \text{, if} \quad |e(t)| < \lambda. \end{cases}$$

Observe that

$$\dot{k}(t) = \theta(t)e(t) = |\theta(t)||e(t)| \quad \text{and} \quad d_\lambda(e(t)) = |\theta(t)| \leq \lambda^{-1}|\theta(t)||e(t)|.$$

Therefore, using the first differential equation in (5.15), we have, for all $t \in [0,\omega)$,

$$\begin{aligned}
\frac{d}{dt}V_\lambda(e(t)) &\leq [A_1 - N(k(t))cb]|\theta(t)||e(t)| + |\theta(t)|\|g_1(\cdot)\|_{L_\infty(0,\omega)} \\
&\quad + \|A_2\|\|\theta(t)\|\|z(t)\| \\
&\leq [M_1 - N(k(t))cb]d_\lambda(e(t))|e(t)| + \|A_2\|\|\theta(t)\|\|z(t)\| \\
&\leq [M_1 - N(k(t))cb]\dot{k}(t) + \|A_2\|\|\theta(t)\|\|z(t)\|, \quad (5.16)
\end{aligned}$$

where $M_1 := |A_1| + \lambda^{-1}\|g_1(\cdot)\|_{L_\infty(0,\omega)}$. Rearranging the second equation in (5.15) as

$$\dot{z}(t) = A_4 z(t) + A_3\theta(t) + h(t),$$

with

$$h(\cdot) := A_3[e(\cdot) - \theta(\cdot)] + g_2(\cdot) \in L_\infty(0,\omega),$$

an application of Lemma 5.2.1 yields, for some $M > 0$ and $M_2 := M\left[1 + \lambda^{-1}\right]$,

$$\begin{aligned}
\int_0^t |\theta(s)|\|z(s)\|ds &\leq M\int_0^t [|\theta(s)| + \theta(s)^2]ds \\
&\leq M\int_0^t [1 + \lambda^{-1}]|\theta(s)||e(s)|ds \\
&\leq M_2\int_0^t d_\lambda(e(s))|e(s)|ds \\
&= M_2[k(t) - k(0)],
\end{aligned}$$

which is valid for all $t \in [0,\omega)$. Integration of $V_\lambda(e(s))$ over $[0,t]$ therefore gives

$$\begin{aligned}
V_\lambda(e(t)) &\leq V_\lambda(e(0)) + \int_0^t [M_1 - N(k(s))cb]\dot{k}(s)ds + \|A_2\|M_2[k(t) - k(0)] \\
&\leq V_\lambda(e(0)) - cb\int_{k(0)}^{k(t)} N(\tau)d\tau + M_3[k(t) - k(0)], \quad (5.17)
\end{aligned}$$

where $M_3 := M_1 + \|A_2\|M_2$.

(d):   If $k(\cdot) \notin L_\infty(0,\omega)$, then (5.17) yields, for $k(t) > k(0)$,

$$V_\lambda(e(t)) \le V_\lambda(e(0)) + [k(t) - k(0)]\left[M_3 - \frac{cb}{k(t) - k(0)}\int_{k(0)}^{k(t)} N(\tau)d\tau\right],$$

and, by the properties of the Nussbaum function $N(\cdot)$, the right hand side becomes negative, thus contradicting the non-negativeness of $V_\lambda(e)$. Therefore $k(\cdot)$ is bounded, which, by (5.17), implies boundedness of $V_\lambda(e)$, and hence $e(\cdot) \in L_\infty(0,\omega)$. By the second equation in (5.15) and since A_4 is asymptotically stable, we may conclude boundedness of $z(\cdot)$ on $[0,\omega)$. If $\omega < \infty$ and $N(\cdot)$ is not continuous but piecewise right continuous, then the solution can be extended to the right in a similar manner as in part (e) of the proof of Theorem 4.2.1. Thus boundedness of $(x(\cdot), k(\cdot))$ yields $\omega = \infty$. This establishes assertions (i)-(iii).

(e):   It remains to show (iv). Note that

$$|\theta(t)|\|z(t)\| \le \lambda^{-1}|\theta(t)||e(t)|\|z(t)\| = \lambda^{-1}k(t)\|z(t)\|,$$

and therefore the boundedness of $z(\cdot)$ yields, for some $M_4 > 0$,

$$|\theta(t)|\|z(t)\| \le M_4\dot{k}(t) \quad \text{for all} \quad t \in [0,\omega).$$

Defining $M_5 := \max_{t\ge0}\{M_1 - N(k(t))cb + \|A_2\|M_4\}$ then, by (5.16),

$$\frac{d}{dt}V_\lambda(e(t)) \le -\dot{k}(t) + [M_5 + 1]\dot{k}(t).$$

Therefore, the derivative of the sign-indefinite Lyapunov function

$$W(\cdot,\cdot) : \mathbb{R}^{m+1} \to \mathbb{R}, \quad (e,k) \mapsto V_\lambda(e) - (M_5 + 1)k$$

along the solution component $e(t)$ and $k(t)$ of (5.15) is, for all $t \ge 0$,

$$\frac{d}{dt}W(e(t), k(t)) \le -\dot{k}(t) = d_\lambda(e(t))|e(t)| \le 0.$$

Now LaSalle's Invariance Principle for non-autonomous systems, see LaSalle (1976), proves that the ω-limit set of the bounded solution $(e(\cdot), x(\cdot), k(\cdot))$ is contained in $\{(e,x,k) \in \mathbb{R}^{n+2} \mid |e| \le \lambda\}$. This proves (iv) and completes the proof. (For a direct proof see part (e) of the proof of Theorem 6.3.2.)   □

Observe that the gain adaptation in Theorem 5.2.2 is very close to that in Theorem 4.2.1: for $\lambda = 0$ and $y_{\text{ref}}(\cdot) \equiv 0$, we rediscover the usual adaptation $\dot{k}(t) = y(t)^2$. However the proof does not extend in any obvious manner to the present setting. Although Theorem 5.2.2 is subsumed by the general results on multivariable systems in Theorem 5.2.4 and 5.2.5, a separate proof has been presented in order to illustrate the difference to the proof of Theorem 4.2.1 and the differences between the single-input, single-output and the multivariable case for λ-tracking.

To prove λ-tracking for multivariable systems another technical lemma is needed.

Lemma 5.2.3

Let $\rho > 0$. For a positive-definite $P = P^T \in \mathbb{R}^{m \times m}$ define the following 'distance' function

$$D_\rho(\cdot) : \mathbb{R}^m \to [0, \infty), \quad e \mapsto \begin{cases} \|e\|_P - \rho & \text{, if } \|e\|_P \geq \rho \\ 0 & \text{, if } \|e\|_P < \rho. \end{cases} \tag{5.18}$$

If, for $p, q > 0$, we have

$$p\|e\| \leq \|e\|_P \leq q\|e\| \quad \text{for all} \quad e \in \mathbb{R}^m,$$

then, for all $e \in \mathbb{R}^m$, the following inequalities are valid

(i) $D_{p\lambda}(e) \geq pd_\lambda(e)$

(ii) $D_{q\lambda}(e) \leq qd_\lambda(e)$.

Proof: Let $\bar{E}_\rho(0)$ denote the closed ellipsoid $\{\xi \,|\, \|\xi\|_P \leq \rho\} \subset \mathbb{R}^m$ and define

$$\hat{k} := \begin{cases} p\lambda\|\xi\|_P^{-1}\xi & \text{, if } \|\xi\|_P \geq p\lambda \\ \xi & \text{, if } \|\xi\|_P < p\lambda \end{cases} \quad \text{and} \quad \tilde{k} := \begin{cases} \lambda\|\xi\|^{-1}\xi & \text{, if } \|\xi\| > \lambda \\ \xi & \text{, if } \|\xi\| \leq \lambda. \end{cases}$$

Then, since $\hat{k} \in \bar{E}_{p\lambda}(0)$ and $\bar{E}_{p\lambda}(0) \subset \bar{B}_\lambda(0)$,

$$\begin{aligned} D_{p\lambda}(\xi) = \|\xi - \hat{k}\|_P \geq p\|\xi - \hat{k}\| &\geq p \min\{\|\xi - k\| \,|\, k \in \bar{E}_{p\lambda}(0)\} \\ &\geq p \min\{\|\xi - k\| \,|\, k \in \bar{B}_\lambda(0)\} \\ &= pd_\lambda(\xi), \end{aligned}$$

and

$$\begin{aligned} qd_\lambda(\xi) = q\|\xi - \tilde{k}\| \geq \|\xi - \tilde{k}\|_P &\geq \min\{\|\xi - k\|_P \,|\, k \in \bar{B}_\lambda(0)\} \\ &\geq \min\{\|\xi - k\|_P \,|\, k \in \bar{E}_{q\lambda}(0)\} \\ &= D_{q\lambda}(\xi). \end{aligned}$$

This completes the proof.   □

Theorem 5.2.4

Suppose

$$\left. \begin{aligned} \dot{x}(t) &= Ax(t) + Bu(t), \quad x(0) = x_0 \\ y(t) &= Cx(t) \end{aligned} \right\} \tag{5.19}$$

with $(A, B, C) \in \mathbb{R}^{n \times n} \times \mathbb{R}^{n \times m} \times \mathbb{R}^{m \times n}$ is minimum phase and $\sigma(CB) \subset \mathbb{C}_+$. Let $\lambda > 0$. If the control strategy

$$\left.\begin{array}{rcl}
e(t) & = & y(t) - y_{\text{ref}}(t), \\
u(t) & = & -k(t)e(t), \\
\dot{k}(t) & = & d_\lambda(e(t))\|e(t)\|, \quad k(0) = k_0
\end{array}\right\} \tag{5.20}$$

is applied to (5.19), for arbitrary $x_0 \in \mathbb{R}^n$, $k_0 \in \mathbb{R}$, $y_{\text{ref}}(\cdot) \in W^{1,\infty}(\mathbb{R}, \mathbb{R}^m)$, then the closed-loop system (5.19), (5.20) has the properties

(i) there exists a unique solution $(x(\cdot), k(\cdot)) : [0, \infty) \to \mathbb{R}^{n+1}$,

(ii) $\lim_{t \to \infty} k(t) = k_\infty$ exists and is finite,

(iii) $x(\cdot), k(\cdot) \in L_\infty(0, \infty)$,

(iv) the error $e(t)$ approaches the closed ball $\bar{B}_\lambda(0)$ as $t \to \infty$.

Proof: As in the proof of Theorem 5.2.2, we may assume that the solution $(x(\cdot), k(\cdot))$ of the closed-loop system (5.19), (5.20) exists on a maximal interval $[0, \omega)$ for some $\omega > 0$. By Lemma 2.1.3, the closed-loop system may be described as

$$\left.\begin{array}{rcl}
\dot{e}(t) = [A_1 - k(t)CB]e(t) & +A_2 z(t) + g_1(t) \,, & e(0) = Cx_0 - y_{\text{ref}}(0) \\
\dot{z}(t) = \qquad\qquad A_3 e(t) & +A_4 z(t) + g_2(t) \,, & z(0) = Nx_0 \\
\dot{k}(t) = \quad d_\lambda(e(t))\|e(t)\| & & , k(0) = k_0
\end{array}\right\} \tag{5.21}$$

where

$$\begin{array}{rcl}
g_1(t) & := & -\dot{y}_{\text{ref}}(t), \\
g_2(t) & := & A_1 y_{\text{ref}}(t) - \dot{y}_{\text{ref}}(t).
\end{array}$$

(a):   For $p := \sqrt{\mu_{\min}(P)}$ and $q := \sqrt{\mu_{\max}(P)} = \|P^{1/2}\|$, we use the notation of Lemma 5.2.3 and define, for $\rho > 0$, the C^1-function

$$V_\rho(\cdot) : \mathbb{R}^m \to [0, \infty), \quad e \mapsto \tfrac{1}{2}D_\rho(e)^2.$$

Differentiation of $V_\rho(\cdot)$ along the solution component $e(t)$ yields, since $\|g_1(\cdot)\|_{L_\infty(0,\omega)} \leq M_1$ for some $M_1 > 0$,

$$\begin{aligned}
\frac{d}{dt} V_\rho(e(t)) \;=\;& D_\rho(e(t))\|e(t)\|_P^{-1} \langle e(t), P\dot{e}(t)\rangle \\
\leq\;& D_\rho(e(t))\|e(t)\|_P^{-1} \left[\|PA_1\|\|e(t)\|^2 + \|PA_2\|\|e(t)\|\|z(t)\| \right. \\
& \left. \qquad\qquad\qquad\qquad\qquad\qquad +M_1\|P\|\|e(t)\|\right] \\
& -k(t)D_\rho(e(t))\|e(t)\|_P^{-1}\langle e(t), PCBe(t)\rangle \\
\leq\;& D_\rho(e(t)) \left[p^{-1}\|PA_1\|\|e(t)\| + p^{-1}\|PA_2\|\|z(t)\| + M_1 q^2 \rho^{-1}\|e(t)\|\right] \\
& -k(t)D_\rho(e(t))\|e(t)\|_P^{-1}\langle e(t), PCBe(t)\rangle \\
\leq\;& M_2 D_\rho(e(t)) \left[\|e(t)\| + \|z(t)\|\right] \\
& -k(t)D_\rho(e(t))\|e(t)\|_P^{-1}\langle e(t), PCBe(t)\rangle \tag{5.22}
\end{aligned}$$

where $M_2 := p^{-1} [\|PA_1\| + \|PA_2\|] + M_1 q^2 \rho^{-1}$.

(b):   If $P \in \mathbb{R}^{m \times m}$ denotes the positive-definite solution of

$$PCB + (CB)^T P = 2I_m,$$

and $t \in [t_0, \omega)$, such that $k(t_0) \geq 0$, then, by (5.22),

$$\frac{d}{dt} V_\rho(e(t)) \leq M_2 D_\rho(e(t)) [\|e(t)\| + \|z(t)\|] - k(t) D_\rho(e(t)) p^{-1} \|e(t)\|. \quad (5.23)$$

Define the continuous map

$$\theta_\rho(\cdot) : \mathbb{R}^m \to [0, \infty), \quad t \mapsto \begin{cases} D_\rho(e(t)) \|e(t)\|_P^{-1} e(t) & , \text{if } \|e(t)\|_P > \rho \\ 0 & , \text{if } \|e(t)\|_P \leq \rho. \end{cases}$$

We remark, in passing, that $\|\theta_\rho(t)\|_P = \|P^{\frac{1}{2}} \theta_\rho(t)\| = D_\rho(e(t))$ and so

$$p\|\theta_\rho(t)\| \leq D_\rho(e(t)) \leq q\|\theta_\rho(t)\|,$$

and

$$D_{p\lambda}(e) \geq 0 \iff \|e\|_P \geq p\lambda \implies \|e\| \geq q^{-1}p\lambda.$$

Since

$$h(\cdot) := A_3[e(\cdot) - \theta_{p\lambda}(\cdot)] + g_2(\cdot) \in L_\infty(0, \omega),$$

an application of Lemma 5.2.1 yields, for some $M > 0$, for all $t \in [t_0, \omega)$ and $M_3 := Mp^{-2}q [1 + \lambda^{-1}]$,

$$\begin{aligned}
\int_{t_0}^t \|\theta_{p\lambda}(s)\| \|z(s)\| ds &\leq M \int_{t_0}^t [\|\theta_{p\lambda}(s)\| + \|\theta_{p\lambda}(s)\|^2] ds \\
&\leq Mp^{-2} \int_{t_0}^t D_{p\lambda}(e(s))[p + D_{p\lambda}(e(s))] ds \\
&\leq Mp^{-2} \int_{t_0}^t D_{p\lambda}(e(s))[p + \|e(s)\|_P] ds \\
&\leq Mp^{-2} \int_{t_0}^t D_{p\lambda}(e(s))[p + q\|e(s)\|] ds \\
&\leq M_3 \int_{t_0}^t D_{p\lambda}(e(s))\|e(s)\| ds.
\end{aligned}$$

Integration of $\frac{d}{ds} V_\zeta(e(s))$ over $[t_0, t]$, $0 \leq t_0 \leq t < \omega$, gives

$$V_{p\lambda}(e(t)) \leq V_{p\lambda}(e(t_0)) + \int_{t_0}^t [M_2 + M_3 - p^{-1}k(s)] D_{p\lambda}(e(s)) \|e(s)\| ds. \quad (5.24)$$

(c):   If $k(\cdot) \notin L_\infty(0, \omega)$, then there exists $t_1 \in [0, \omega)$ so that

$$M_2 + M_3 - q^{-1}k(s) < 0 \quad \text{for all} \quad s \in [t_1, \omega)$$

and, by Lemma 5.2.3 (i) and (5.24), we have

$$
\begin{aligned}
V_{p\lambda}(e(t)) &\le V_{p\lambda}(e(t_1)) + \int_{t_1}^{t} p[M_2 + M_3 - p^{-1}k(s)]d_\lambda(e(s))\|e(s)\|ds \\
&= V_{p\lambda}(e(t_1)) + \int_{k(t_1)}^{k(t)} p[M_2 + M_3 - p^{-1}\tau]d\tau.
\end{aligned}
$$

Since $\lim_{t\to\omega} k(t) = \infty$, the right hand side takes nagative values, thus contradicting the non-negativeness of the left hand side. Similar to part (d) of the proof of Theorem 5.2.2 the assertions (i)-(iii) follow.

(d): It remains to prove (iv). From (5.22), boundedness of $z(\cdot)$ and $k(\cdot)$, and Lemma 5.2.3 (ii) we deduce, for some $M_4 > 0$,

$$
\frac{d}{dt}V_{q\lambda}(e(t)) \le M_4 D_{q\lambda}(e(t))\|e(t)\| \le M_4 q\dot{k}(t).
$$

Therefore, the derivative of the C^1-function

$$
W(\cdot) : \mathbb{R}^{m+1} \to \mathbb{R}, \quad (e, k) \mapsto V_{q\lambda}(e) - [M_4 q + 1]\,k
$$

along the solution components e and k of (5.21) satisfies

$$
\frac{d}{dt}W(e(t), k(t)) \le -\dot{k}(t) = -d_\lambda(e(t))\|e(t)\| \le 0.
$$

Now the remainder of the proof follows as in part (e) of the proof of Theorem 5.2.2: This completes the proof. □

Unfortunately, we are not able to extend Theorem 5.2.4 to the system class of multivariable systems (A, B, C) where it is only known that $\sigma(CB) \subset \mathbb{C}_+$ or $\sigma(CB) \subset \mathbb{C}_-$, but unknown in which complex half plane the spectrum is lying. However, the following weaker result is achieved.

Theorem 5.2.5
Suppose

$$
\left.\begin{aligned}
\dot{x}(t) &= Ax(t) + Bu(t), \quad x(0) = x_0 \\
y(t) &= Cx(t),
\end{aligned}\right\} \tag{5.25}
$$

with $(A, B, C) \in \mathbb{R}^{n\times n} \times \mathbb{R}^{n\times m} \times \mathbb{R}^{m\times n}$, is minimum phase and there exist positive-definite $P = P^T, Q = Q^T \in \mathbb{R}^{m\times m}$ and $\beta \in \{-1, +1\}$ so that

$$
PCB + (CB)^T P = 2\beta Q.
$$

If $p > 0$ is known so that

$$
p\|e\| \le \|e\|_P \quad \text{for all} \quad e \in \mathbb{R}^m,
$$

then for every scaling-invariant Nussbaum function $N(\cdot) : \mathbb{R} \rightarrow \mathbb{R}$ and arbitrary $\lambda > 0$, the control strategy

$$
\left.
\begin{aligned}
e(t) &= y(t) - y_{\text{ref}}(t), \\
u(t) &= -N(k(t))e(t), \\
\dot{k}(t) &= D_{p\lambda}(e(t))\|e(t)\|, \quad k(0) = k_0,
\end{aligned}
\right\} \tag{5.26}
$$

where the notation (5.18) is used, applied to (5.25), for arbitrary $x_0 \in \mathbb{R}^n$, $k_0 \in \mathbb{R}$, $y_{\text{ref}}(\cdot) \in \mathcal{W}^{1,\infty}(\mathbb{R}, \mathbb{R}^m)$, yields the closed-loop system (5.25), (5.26) with the properties

(i) there exists a unique solution $(x(\cdot), k(\cdot)) : [0, \infty) \rightarrow \mathbb{R}^{n+1}$,

(ii) $\lim_{t \rightarrow \infty} k(t) = k_\infty$ exists and is finite,

(iii) $x(\cdot), k(\cdot) \in L_\infty(0, \infty)$,

(iv) the error $e(t)$ approaches the closed ball $\bar{B}_\lambda(0)$ as $t \rightarrow \infty$.

Proof: As in the proof of Theorem 5.2.4, we may assume that the solution $(x(\cdot), k(\cdot))$ exists on a maximal interval $[0, \omega)$ for some $\omega > 0$ and the closed-loop system may be described by (5.21) with d_λ replaced by $D_{p\lambda}$ and k replaced by $N(k)$.

(a):   Set $q = \|P^{1/2}\|$. Since

$$
D_{p\lambda}(e) > 0 \quad \Longrightarrow \quad p\lambda < \|e\|_P \le q\|e\|,
$$

we have

$$
\begin{aligned}
-N(k(t))D_{p\lambda}(e(t))\frac{\langle e(t), PCBe(t)\rangle}{\|e(t)\|_P} &= -\beta N(k(t))D_{p\lambda}(e(t))\|e(t)\|_P^{-1}\|e(t)\|_Q^2 \\
&\le -\tilde{N}(k(t))D_{p\lambda}(e(t))\|e(t)\|,
\end{aligned}
$$

where

$$
\tilde{N}(k) := \begin{cases} -\beta p^{-1}\|Q\|N(k) & \text{, if } \beta N(k) \le 0 \\ -\beta q^{-1}\mu_{\min}(Q)^2 N(k) & \text{, if } \beta N(k) > 0. \end{cases}
$$

By Lemma 5.2.1 and arguing as in part (b) of the proof of Theorem 5.2.4, there exists $M_5 > 0$ such that

$$
\int_0^t \|\theta_{p\lambda}(s)\|\|z(s)\|ds \le M_5 \int_0^t D_{p\lambda}(e(s))\|e(s)\|ds
$$

Therefore, by (5.22) with k replaced by $N(k)$, we have

$$
\begin{aligned}
V_{p\lambda}(e(t)) \le\; & V_{p\lambda}(e(t_0)) + [M_2 + M_5]\int_0^t D_{p\lambda}(e(s))\|e(s)\|ds \\
& -\int_0^t \tilde{N}(k(s))D_{p\lambda}(e(s))\|e(s)\|ds.
\end{aligned}
$$

Since $\tilde{N}(\cdot)$ is a Nussbaum function too, assertions (i)-(iii) follow as in part (d) of the proof of Theorem 5.2.2.

(b): It remains to prove (iv). Since $e(\cdot)$, $z(\cdot)$, and $k(\cdot)$ are bounded, we conclude from (5.22), with k replaced by $N(k)$, for some $M_5 > 0$,

$$\frac{d}{dt}V_{p\lambda}(e(t)) \leq M_5 D_{p\lambda}(e(t))\|e(t)\| = M_5\dot{k}(t)$$

and the remainder of the proof follows as in part (d) and (e) of the proof of Theorem 5.2.2.

This completes the proof. □

Remark 5.2.6
It is easy to see that the (piecewise) continuous feedback strategy in Theorem 5.2.4 and 5.2.5 can be replaced by a piecewise constant gain implementation as suggested in Theorem 4.5.1 with $y(t)$ replaced by $e(t)$. In case of Theorem 5.2.5 and $m > 1$ it has to be assumed that $\{T_i\}_{i \in \mathbb{N}}$ is scaling-invariant as defined in Remark 4.5.2 (i).

5.3 Notes and References

The use of an internal model to solve the universal adaptive tracking problem goes back to Mareels (1984) for single-input, single-output systems of relative degree $p \geq 1$ with known upper bound for the positive high-frequency gain. Helmke et al. (1990) proved Theorem 5.1.3 for single-input, single-output, minimum phase systems (A, b, c) with $cb \neq 0$ and unknown order of the state dimension.

The generalization for multivariable, minimum phase system (A, B, C) with $\det CB \neq 0$ has been proved by: Townley and Owens (1991) for $\alpha(\cdot)$ with zeros all of which lie on the imaginary axis; Miller and Davison (1991a) for $\alpha(\cdot)$ having only roots in $\mathbb{C}_+$; Logemann and Ilchmann (1991) for arbitrary $\alpha(\cdot)$. All proof are different and were found independently. The result by Logemann and Ilchmann (1991) covers a large class of infinite-dimensional systems.

Lemma 5.1.2 is from Miller and Davison (1991a), slightly extended for systems (A, B, C, D) with $D + D^T \geq 0$.

All results of Section 5.2 are from Ilchmann and Ryan (1992).

Chapter 6

Robustness

In this section, we study well posedness and robustness properties of the universal adaptive controllers and tracking controllers introduced in Chapter 4 and 5. In Section 6.1, it will be shown that all stabilizers can cope with nonlinear, time-varying, additive input-output perturbations, provided they are linearly bounded and, depending on where the system is perturbed, the bound has to be sufficiently small (in terms of the system entries). An arbitrary additive signal $d(\cdot) \in L_p(0, \infty)$ in the state equation is also tolerated. See Figure 6.1.

In Section 6.2, we shall prove that many universal adaptive stabilizers of single-input, single-output, (but also certain multivariable) systems tolerate sector bounded nonlinearities in the actuator and sensor. The bounds of the nonlinearities need not to be known.

In Section 6.3, it will be proved that the λ-tracking controller, introduced in Section 5.2, is robust with respect to much larger classes of nonlinearities than considered in Section 6.1. The perturbation results are also remarkable for λ-stabilization only. The asymptotic tracking controller involving an internal model, see Section 5.1, is not well posed or robust with respect to nonlinearities as considered in Section 6.1 and 6.2. The λ-tracking controller overcomes this disadvantage and, furthermore, copes with additive noise corrupted input and output which is not possible for asymptotic stabilization.

6.1 Additive nonlinear state and input perturbations

In this section, we will consider various universal adaptive stabilizers when applied to minimum phase systems $(A, B, C, D) \in \mathbb{R}^{n \times n} \times \mathbb{R}^{n \times m} \times \mathbb{R}^{m \times n} \times$

$\mathbb{R}^{m \times m}$ which are now subjected to perturbations of the form

$$
\left.
\begin{aligned}
\dot{x}(t) &= Ax(t) + g_1(t, x(t)) + g_2(t, y(t)) + d(t) \\
&\quad + B[u(t) + h(t, u(t))] \\
y(t) &= Cx(t) + Du(t).
\end{aligned}
\right\} \quad , x(0) = x_0 \qquad (6.1)
$$

Throughout this section, $d(\cdot) \in L_p(\mathbb{R})$, where $p \geq 1$ will be specified later, and the nonlinearities

$$
\begin{array}{llll}
g_1(\cdot, \cdot) & : & \mathbb{R} \times \mathbb{R}^n \to \mathbb{R}^n & , \|g_1(t, x)\| \leq \hat{g}_1 \|x\| \\
g_2(\cdot, \cdot) & : & \mathbb{R} \times \mathbb{R}^m \to \mathbb{R}^m & , \|g_2(t, y)\| \leq \hat{g}_2 \|y\| \\
h(\cdot, \cdot) & : & \mathbb{R} \times \mathbb{R}^m \to \mathbb{R}^m & , \|h(t, u)\| \leq \hat{h} \|u\|
\end{array}
$$

are assumed to be *Carathéodory functions* [1], which, for some unknown $\hat{g}_1, \hat{g}_2, \hat{h} \geq 0$, are linearly bounded for almost all $t \in \mathbb{R}$ and for all $x \in \mathbb{R}^n$, $u, y \in \mathbb{R}^m$.

These assumptions ensure that for every locally integrable $u(\cdot)$, and every $x_0 \in \mathbb{R}^n$, the initial value problem (6.1) possesses a solution $x(\cdot) : [0, \omega) \to \mathbb{R}^n$, maximally extended over $[0, \omega)$, for some $\omega > 0$. $x(\cdot)$ is a solution in the sense that it is absolutely continuous on compact intervals and satisfies the initial value problem for almost all $t \in [0, \omega)$.

If further assumptions are made: $x \mapsto g_1(t, x)$, $y \mapsto g_2(t, y)$ are locally Lipschitz for each fixed $t \in \mathbb{R}$, and $t \mapsto g_1(t, x)$, $t \mapsto g_2(t, y)$, $t \mapsto h(t, u)$ are locally integrable for each fixed x, y, respectively u, then uniqueness of the solution is guaranteed.

However, uniqueness is not important for our purposes because it will be shown that, if a universal adaptive stabilizer is applied, *every* solution of the closed-loop system meets the desired control objectives.

We shall see that the nonlinearities $g_1(t, x)$ and $h(t, u)$ show well posedness instead of robustness, that means, previous adaptive stabilizers are applicable if the linear bounds $\hat{g}_1$, $\hat{h}$ are *sufficiently small*, in terms of the system entries (A, B, C, D). There is no restriction on the linear bound of $g_2(t, y)$, and $d(\cdot)$ is an arbitrary L_p-function, where p corresponds to the p used in the gain adaptation $\dot{k}(t) = \|y(t)\|^p$.

The stability proofs in Chapter 4 were based on the inequalities in Lemma 2.1.6 and Lemma 3.1.6. First we shall establish similar inequalities taking into account the nonlinearities.

---

[1] $f : \mathbb{R} \times \mathbb{R}^q \to \mathbb{R}$ is called a Carathéodory function, if $f(\cdot, x) : t \mapsto f(t, x)$ is measurable on $\mathbb{R}$ for each $x \in \mathbb{R}^q$, and $f(t, \cdot) : x \mapsto f(t, x)$ is continuous on $\mathbb{R}^q$ for all $t \in \mathbb{R}$.

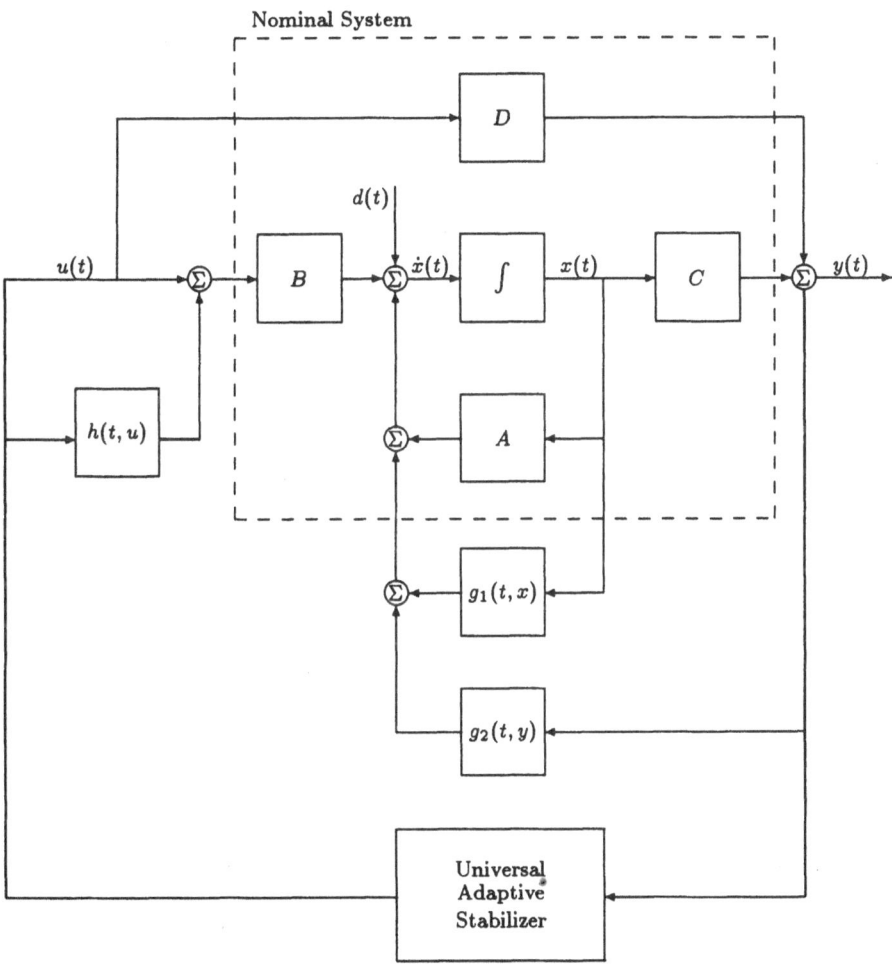

Figure 6.1: Adaptive stabilization of nonlinearly perturbed systems

Lemma 6.1.1
Consider the initial value problem (6.1) with $D = 0$ and some locally integrable
$u(\cdot) : [0, \infty) \rightarrow \mathbb{R}^m$. Let $x(\cdot) : [0, \omega) \rightarrow \mathbb{R}^n$ be an absolutely continuous
solution, for some $\omega > 0$, and let $p \geq 1$, $P = P^T \in \mathbb{R}^{m \times m}$ positive-definite,
and $(A, B, C) \in \mathbb{R}^{n \times n} \times \mathbb{R}^{n \times m} \times \mathbb{R}^{m \times n}$ be minimum phase. If the linear bound
$\hat{g}_1$ is sufficiently small, depending on the entries of (A, B, C), then there exists
an $M > 0$ such that the output of (6.1) satisfies, for all $t \in [0, \omega)$,

$$\frac{1}{p}\|y(t)\|_P^p \leq M + M\int_0^t \|y(s)\|^p \, ds + \int_0^t \|y(s)\|_P^{p-1} \langle \beta(y(s)), P\, CB[u(s)+h(s,u(s))]\rangle) ds$$

where

$$\beta(\cdot) : \mathbb{R}^m \rightarrow \mathbb{R}^m, \qquad y \mapsto \beta(y) = \begin{cases} \frac{y}{\|y\|_P} &, \quad \text{if} \quad y \neq 0 \\ 0 &, \quad \text{if} \quad y = 0. \end{cases}$$

Proof: Applying the coordinate transformation $S^{-1}x = (y^T, z^T)^T$ of
Lemma 2.1.3 to (6.1) yields

$$\left.\begin{aligned} \dot{y}(t) &= A_1 y(t) + A_2 z(t) + C\left[g_1(t, x(t)) + g_2(t, y(t)) + d(t)\right] \\ &\quad + CB[u(t) + h(s, u(t))] \\ \dot{z}(t) &= A_3 y(t) + A_4 z(t) + N\left[g_1(t, x(t)) + g_2(t, y(t)) + d(t)\right]. \end{aligned}\right\} \quad (6.2)$$

Since A_4 is exponentially stable and

$$\|N\left[g_1(t, x) + g_2(t, y)\right]\| \leq \|N\| \left[\hat{g}_1\|S\|(\|y\| + \|z\|) + \hat{g}_2\|y\|\right]$$

for almost all $t \in \mathbb{R}$ and all $(x, y) \in \mathbb{R}^n \times \mathbb{R}^m$, there exist $M_1, \lambda > 0$ such that
Variation of Constants applied to the second equation in (6.2) yields

$$\|z(t)\| \leq M_1 e^{-\lambda t}\|Nx_0\| + M_1 \mathcal{L}\left(\hat{g}_1\|z(\cdot)\| + \|y(\cdot)\| + \|d(\cdot)\|\right)(t),$$

where we have used the bounded operator

$$\mathcal{L} : L_p(\mathbb{R}_+) \rightarrow L_p(\mathbb{R}_+), \quad \zeta(\cdot) \mapsto \left(t \mapsto \mathcal{L}\left(\zeta(\cdot)\right)(t) := \int_0^t e^{-\lambda(t-s)}\zeta(s)ds\right).$$

Taking L_p-norms in the above inequality and using

$$\|\mathcal{L}\left(\|\zeta(\cdot)\|\right)(\cdot)\|_{L_p(0,t)} \leq \lambda^{-1}\|\zeta(\cdot)\|_{L_p(0,t)},$$

see Theorem 6.5.54 in Vidyasagar (1978), yields

$$\|z(\cdot)\|_{L_p(0,t)} \leq \lambda^{-1/p} M_1 \|Nx_0\|$$
$$+ M_1 \lambda^{-1}\left[\hat{g}_1\|z(\cdot)\|_{L_p(0,t)} + \|y(\cdot)\|_{L_p(0,t)} + \|d(\cdot)\|_{L_p(0,t)}\right]$$

and hence, for $\hat{g} > 0$ sufficiently small so that $1 - M_1 \lambda^{-1} \hat{g} > 0$, and

$$M_2 := \left(1 - M_1 \lambda^{-1} \hat{g}\right)^{-1} \left[\lambda^{-1/p} M_1 \|N x_0\| + M_1 \lambda^{-1} \left(1 + \|d(\cdot)\|_{L_p(0,\infty)}\right)\right],$$

we have

$$\|z(\cdot)\|_{L_p(0,t)} \leq M_2 + M_2 \|y(\cdot)\|_{L_p(0,t)}. \tag{6.3}$$

Note that $y(\cdot)$ is an absolutely continuous function. Let $J_1 \subset [0, \infty)$ be the set of measure zero where $y(\cdot)$ is not differentiable and

$$J_2 := \{t \in \mathbb{R}_+ \setminus J_1 | \, y(t) = 0 \quad \text{and} \quad \dot{y}(t) \neq 0\}.$$

It is easy to see that $\|y(\cdot)\|_P$ is not differentiable in any point of J_2. However $\|y(\cdot)\|_P$ is absolutely continuous because $y(\cdot)$ is and hence J_2 must be of measure zero. It follows that $J := J_1 \cup J_2$ is of measure zero and a routine calculation gives, for all $s \in \mathbb{R}_+ \setminus J$,

$$\frac{d}{ds}\|y(s)\|_P = \begin{cases} \dfrac{\langle y(s), P\dot{y}(s) \rangle}{\|y(s)\|_P} & , y(s) \neq 0 \\ 0 & , y(s) = 0. \end{cases}$$

Now integration of $\frac{d}{ds}\left(\frac{1}{p}\|y(s)\|_P^p\right)$ over $[0, t]$ yields, for all $t \in [0, \omega)$,

$$\frac{1}{p}\|y(t)\|_P^p \leq \frac{1}{p}\|y(0)\|_P^p + \int_0^t \|y(s)\|_P^{p-1} \langle \beta(y(s)), P\left[A_1 y(s) + A_2 z(s)\right]$$

$$+ PC[g_1(s, x(s)) + g_2(s, y(s)) + d(s)]$$

$$+ PCB[u(s) + h(s, u(s))]\rangle \, ds$$

$$\leq M_3 + M_3 \left[\|y(\cdot)\|_{L_p(0,t)}^p + \int_0^t \|y(s)\|_P^{p-1}[\|z(s)\| + \|d(s)\|]ds\right]$$

$$+ \int_0^t \|y(s)\|_P^{p-1} \langle \beta(y(s)), PCB[u(s) + h(s, u(s))]\rangle \, ds,$$

where

$$M_3 := \frac{1}{p}\|y(0)\|_P^p + \|P\|^{\frac{p-2}{2}} \left[\|PA_1\| + \|PA_2\| + \|PC\|(2\hat{g}_1\|S\| + \hat{g}_2)\right],$$

and hence, by (2.11) and (6.3),

$$\frac{1}{p}\|y(t)\|_P^p \leq M_3 + M_3(1 + M_2)\|y(\cdot)\|_{L_p(0,t)}^p$$

$$+ M_3 \left(M_2 + \|d(\cdot)\|_{L_p(0,\infty)}\right) \|y(\cdot)\|_{L_p(0,t)}^{p-1}$$

$$+ \int_0^t \|y(s)\|_P^{p-1} \langle \beta(y(s)), PCB[u(s) + h(s, u(s))]\rangle ds$$

$$\leq \ +M_3 \left(M_2 + \|d(\cdot)\|_{L_p(0,\infty)}\right) \left[1 + \|y(\cdot)\|^p_{L_p(0,t)}\right]$$

$$+ \int_0^t \|y(s)\|_P^{p-1} \langle \beta(y(s)), PCB[u(s) + h(s, u(s))]\rangle ds.$$

This proves the statement. $\square$

Lemma 6.1.2

Suppose the assumptions of Lemma 3.1.6 are valid. If instead of (3.12) the perturbed system (6.1) with $h(t, u) \equiv 0$ is considered, and $x(\cdot) : [0, \omega) \to \mathbb{R}^n$ is a solution of the closed-loop system (6.1), $u(t) = -\mathcal{K}(t)y(t)$, for some $\omega > 0$, then the derivative of $V(x) = \langle x, Px\rangle$ along $x(t)$ satisfies, for almost all $t \in [0, \omega)$,

$$\frac{d}{dt}V(x(t)) \ \leq \ -2\left[\mu - \hat{g}_1\|P\|\mu_{\min}(P)^{-1}\right]V(x(t)) - \langle y(t), [\mathcal{K}(t) + \mathcal{K}^T(t)]y(t)\rangle$$

$$+2\|K\|\,\|y(t)\|^2 + 2\|P\|\hat{g}_2\|x(t)\|\,\|y(t)\| + 2\langle x(t), Pd(t)\rangle$$

Proof: Similar to (3.15) we have, for almost all $t \in [0, \omega)$,

$$\dot{x}(t) \ = \ \hat{A}x(t) - \hat{B}(\mathcal{K}(t) - K)y(t) + g_1(t, x(t)) + g_2(t, y(t)) + d(t)$$
$$y(t) \ = \ \hat{C}x(t) - \hat{D}(\mathcal{K}(t) - K)y(t),$$

and hence

$$\frac{d}{dt}V(x(t)) \ = \ 2\langle x(t), P\hat{A}x(t)\rangle - 2\langle x(t), P\hat{B}(\mathcal{K}(t) - K)y(t)\rangle$$

$$+2\langle x(t), P[g_1(t, x(t)) + g_2(t, y(t)) + d(t)]\rangle.$$

Now the claim follows from the proof of Lemma 3.1.6. $\square$

The following theorem shows well posedness of universal adaptive stabilization of a class of linearly perturbed multivariable systems (A, B, C), where the spectrum of CB lies either in the left or right half plane.

Theorem 6.1.3

Suppose the system $(A, B, C) \in \mathbb{R}^{n \times n} \times \mathbb{R}^{n \times m} \times \mathbb{R}^{m \times n}$ is minimum phase and $\sigma(CB) \subset \mathbb{C}_+$ or $\mathbb{C}_-$, but unknown in which half of the complex plane the eigenvalues are lying. If the adaptive feedback strategy

$$\left.\begin{array}{rcl} u(t) & = & -\ln k(t) \cos \sqrt{\ln k(t)}\, y(t) \\ \dot{k}(t) & = & \|y(t)\|^2 \end{array}\right\} , k(0) = k_0 \qquad (6.4)$$

is applied to the perturbed system

$$\begin{aligned}
\dot{x}(t) &= Ax(t) + g_1(t, x(t)) + g_2(t, y(t)) + d(t) \\
&\quad + B[u(t) + h(t, u(t))] \qquad\qquad , x(0) = x_0 \\
y(t) &= Cx(t),
\end{aligned} \right\} \qquad (6.5)$$

and the linear bounds $\hat{g}_1, \hat{h} > 0$ are suffiently small, in terms of (A, B, C), then for any $x_0 \in \mathbb{R}^n$, $d(\cdot) \in L_2(\mathbb{R})$, and $k_0 > 1$, there exists a solution $(x(\cdot), k(\cdot)) : [0, \omega) \rightarrow \mathbb{R}^{n+1}$ of the closed-loop system (6.4), (6.5) for some $\omega > 0$, and every solution has on its maximal interval of existence $[0, \omega)$ the properties

(i) $\omega = \infty$,

(ii) $\lim_{t \to \infty} k(t) = k_\infty$ exists and is finite,

(iii) $x(\cdot) \in L_2(0, \infty) \cap L_\infty(0, \infty)$ and $\lim_{t \to \infty} x(t) = 0$.

Proof: The crucial part of the proof is to show that $k(\cdot) \in L_\infty(0, \omega)$ yields a contradiction. The remainder can be done as in the proof of Theorem 4.2.1 and is omitted. We use the notation

$$N(k) = \ln k \, \cos \sqrt{\ln k}.$$

Let $P = P^T \in \mathbb{R}^{m \times m}$ be the positive-definite solution of

$$PCB + (CB)^T P = 2\sigma I_m$$

for some $\sigma \in \{-1, +1\}$. By monotonicity of $k(s)$ we have $N(k(s)) \leq \ln k(t)$ for all $s \in [0, t]$, and hence, by Lemma 6.1.1, for all $t \in [0, \omega)$,

$$\begin{aligned}
\tfrac{1}{2} \|y(t)\|_P^2 &\leq M + M \|y(\cdot)\|_{L_2(0,t)}^2 \\
&\quad + \hat{h} \|PCB\| \ln k(t) \int_0^t \|y(s)\|^2 ds - \sigma \int_0^t N(k(s)) \dot{k}(s) ds,
\end{aligned}$$

whence, for $k(t) > k(0)$,

$$\tfrac{1}{2} \|y(t)\|_P^2 \leq M + [k(t) - k(0)] \left[M + \hat{h} \|PCB\| \ln k(t) - \frac{\sigma}{k(t) - k(0)} \int_{k(0)}^{k(t)} N(\tau) d\tau \right].$$

If it can be shown that

$$\liminf_{k \to \infty} \left[\hat{h} \|PCB\| \ln k - \frac{\sigma}{k - k(0)} \int_{k(0)}^{k} N(\tau) d\tau \right] = -\infty,$$

then this yields a contradiction, and hence $k(\cdot) \in L_\infty(0, \omega)$ and the proof would be complete.

Without restriction of generality we may assume that $k(0) = 1$. By the proof of Lemma 4.1.3 we have, if $\hat{h} < \|PCB\|^{-1}$,

$$\inf_{k>1}\left[\hat{h}\|PCB\|\ln k - \frac{\sigma}{k-k(0)}\int_{k(0)}^{k} N(\tau)d\tau\right]$$

$$= \inf_{k>1}\left[\hat{h}\|PCB\|\ln k - \sigma\ln k \cos\sqrt{\ln k}\right]$$

$$= \inf_{k>1}\left[\hat{h}\|PCB\| - \sigma\cos\sqrt{\ln k}\right]\ln k$$

$$= -\infty.$$

This completes the proof.   □

Remark 6.1.4

(i) An immediate consequence of Theorem 6.1.3 is that the universal adaptive stabilizer (6.4) tolerates time-varying additive perturbations in the system to be stabilized of the form

$$\dot{x}(t) = \left[A + \tilde{A}(t)\right]x(t) + \left[B + \tilde{B}(t)\right]u(t),$$
$$y(t) = Cx(t),$$

provided $\tilde{A}(t)$ and $\tilde{B}(t)$ are measurable and uniformly bounded with sufficiently small bound. Note that no assumption on the time-variation of $\tilde{A}(t)$ and $\tilde{B}(t)$ has to be made, the derivative may even not exist.

(ii) We are unable to prove Theorem 6.1.3 for $u(t) = -N(k(t))y(t)$, where $N(\cdot)$ is an arbitrary switching function. It seem that the special properties of the Nussbaum function used in Theorem 6.1.3, namely $\lim_{k\to\infty}\frac{d}{dk}N(k) = 0$, is necessary to cope with the perturbation.

Well posedness, as proved in Theorem 6.1.3, goes through for switching strategies involving a piecewise constant gain implementation as introduced in Section 4.5. This will be established in the following theorem.

Theorem 6.1.5
Theorem 6.1.3 is also valid if the feedback strategy is replaced by the one used in Theorem 4.5.1 (β), provided the sequence of thresholds $\{T_i\}_{i\in\mathbb{N}}$ is scaling-invariant.

Proof: Again we only prove that $s(\cdot) \notin L_\infty(0, \omega)$ yields a contradiction, the remainder of the proof proceeds in a similar manner to that of Theorem 4.5.1.

Choosing $\sigma \in \{-1, +1\}$ and positive-definite $P \in \mathbb{R}^{m \times m}$ as in the proof of Theorem 6.1.3, Lemma 6.1.1 yields, for some $M > 0$ and all $t \in [0, \omega)$,

$$\frac{1}{2}\|y(t)\|_P^2 \leq M + M\|y(\cdot)\|_{L_2(0,t)}^2$$

$$+ \int_0^t \left[\hat{h}\|PCB\|\,|N(s(\mu))| - \sigma N(s(\mu)) \right] \|y(\mu)\|^2 d\mu.$$

If $t_1 < t_2 < \ldots$ are defined as in part (a) of the proof of Theorem 4.5.1 and $s(\cdot) \notin L_\infty(0, \omega)$, we conclude

$$\frac{1}{2}\|y(t_l)\|_P^2 \leq M + M\,[T_l - T_0] + \sum_{j=0}^{l} \left(\hat{h}\|PCB\| - \sigma(-1)^i \right) T_i[T_i - T_{i-1}]$$

$$= M + [T_l - T_0] \left[M + \frac{\sum_{j=0}^{l} \left(\hat{h}\|PCB\| - \sigma(-1)^i \right) T_i[T_i - T_{i-1}]}{T_l - T_0} \right].$$

If $\hat{h} < \|PCB\|^{-1}$, then the scaling-invariance property of $\{T_i\}_{i \in \mathbb{N}}$ yields that the right hand side of the above inequality takes arbitrary large negative values as $T_i \to \infty$, thus contradicting the non-negativeness of the left hand side. This proves $s(\cdot) \in L_\infty(0, \omega)$ and the remainder of the proof follows as in the proof of Theorem 4.5.1.   □

Finally, it will be shown, that well posedness respectively robustness with respect to state perturbations is guaranteed for most of the universal adaptive stabilization mechanisms presented in Chapter 4.

Theorem 6.1.6
Theorems 4.2.1, 4.5.1, 4.2.4, 4.2.6, 4.5.5, and 4.4.6 remain valid, apart from uniqueness of the solution, if instead of the system considered in the theorems the nonlinear perturbed system

$$\dot{x}(t) = Ax(t) + g_1(t, x(t)) + g_2(t, y(t)) + d(t) + Bu(t) \quad , x(0) = x_0$$
$$y(t) = Cx(t) + Du(t)$$

is considered, where D might be 0, depending on the system class under consideration in the theorems above, the linear bound $\hat{g}_1$ has to be sufficiently small, and $d(\cdot) \in L_p(0, \infty)$, where p corresponds to the adaptation law $k(t) = \|y(t)\|^p$. If exponential stabilization as in Theorem 4.4.6 is considered, it is assumed that there exists an $\varepsilon > 0$ so that $e^{\varepsilon t} d(t) \in L_p(0, \infty)$.

Proof: The crucial part of the proof is to show that $k(\cdot) \notin L_\infty(0, \omega)$ yields a contradiction. The remainder is straightforward and similar to the proofs of the theorems listed above. It is omitted.

Using the inequality in Lemma 6.1.1, the proof of Theorem 4.2.1 and 4.5.1 for the perturbed system is analogous.

Suppose the assumptions of Theorem 4.2.4 are valid. Denoting the feedback chosen by $u(t) = -\mathcal{K}(t)y(t)$, the inequality in Lemma 6.1.2 gives, for almost all $t \in [0, \omega)$,

$$
\begin{aligned}
\frac{d}{dt}V(x(t)) \leq\ & -\mu'V(x(t)) - \langle y(t), [\mathcal{K}(t) + \mathcal{K}^T(t)]\, y(t)\rangle + 2\|K\|\|y(t)\|^2 \\
& + 2\|P\|\|x(t)\|\|d(t)\| + 2\hat{g}_2\|P\|\|x(t)\|\|y(t)\|,
\end{aligned}
$$

where

$$
\mu' := 2\left(\mu - \hat{g}_1\|P\|\mu_{\min}(P)^{-1}\right).
$$

Choose $\hat{g}_1 > 0$ sufficiently small so that $\mu' > 0$. For $\alpha > 0$ we have

$$
\begin{aligned}
2\|P\|\|x(t)\|\|d(t)\| &\leq\ \|P\|^2\alpha^{-2}\|x(t)\|^2 &+&\ \alpha^2\|d(t)\|^2 \\
2\hat{g}_1\|P\|\|x(t)\|\|y(t)\| &\leq\ (\hat{g}_1\|P\|)^2\alpha^{-2}\|x(t)\|^2 &+&\ \alpha^2\|y(t)\|^2.
\end{aligned}
$$

Now, for $\alpha > 0$ sufficiently large so that

$$
\bar{\mu} := \mu' - \frac{(1 + \hat{g}_1^2)\|P\|^2}{\alpha^2}\mu_{\min}(P)^{-1} > 0,
$$

we have, for almost all $t \in [0, \omega)$,

$$
\begin{aligned}
\frac{d}{dt}V(x(t)) \leq\ & -\bar{\mu}V(x(t)) - \langle y(t), [\mathcal{K}(t) + \mathcal{K}^T(t)]\, y(t)\rangle \\
& + \left(2\|K\| + \alpha^2\right)\|y(t)\|^2 + \alpha^2\|d(t)\|^2.
\end{aligned}
$$

By integration of the inequality and using the assumption that $d(\cdot) \in L_2(0, \infty)$, the proof can be completed analogously to the proof of Theorem 4.2.4.

The proof of Theorem 4.2.6, 4.5.5, and 4.4.6 for the perturbed system uses similar ideas as above and is omitted. $\square$

6.2   Sector bounded input-output nonlinearities

The goal of this section is to show that previously introduced universal adaptive stabilizers can tolerate sector bounded nonlinearities in the input and output. It is possible to combine the results of Section 6.1 with the results of this section. However, for the sake of a clearer presentation, it is preferred to describe input-ouput nonlinearities separately. We will consider various classes of systems of the form

$$
\begin{aligned}
\dot{x}(t) &= Ax(t) + B\xi(t, u(t)) \quad , x(0) = x_0 \\
y(t) &= Cx(t),
\end{aligned}
$$

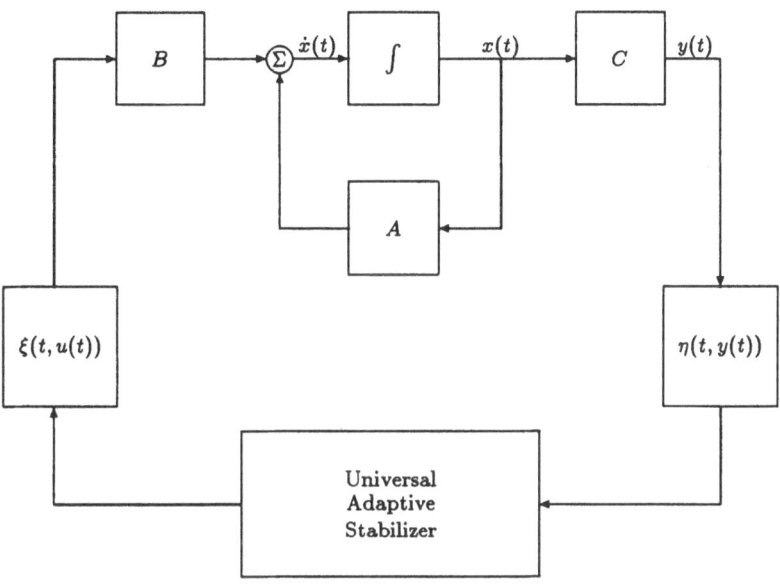

Figure 6.2: Adaptive stabilization in the presence of nonlinear sector bounded input-output nonlinearities.

with $(A, B, C) \in \mathbb{R}^{n \times n} \times \mathbb{R}^{n \times m} \times \mathbb{R}^{m \times n}$ minimum phase, where $\xi(t, u(t))$ represents a time-varying actuator nonlinearity and the output may also be not directly available but via $\eta(t, y(t))$, a time-varying sensor nonlinearity.

We first consider single-input, single-putput systems and assume that

$$\left.\begin{array}{llll} \xi(\cdot, \cdot): \mathbb{R} \times \mathbb{R} \to \mathbb{R} & , \; \xi_1 u^2 & \leq & \xi(t, u)u & \leq & \xi_2 u^2 \\ \eta(\cdot, \cdot): \mathbb{R} \times \mathbb{R} \to \mathbb{R} & , \; \eta_1 y^2 & \leq & \eta(t, y)y & \leq & \eta_2 y^2 \end{array}\right\} \quad (6.6)$$

are sector bounded Carathéodory functions, so that the inequalities in (6.6) hold for some (unknown) $0 < \xi_1 < \xi_2, 0 < \eta_1 < \eta_2$, for almost all $t \in \mathbb{R}$, and for all $u, y \in \mathbb{R}$.

The following theorem shows that the adaptation mechanism introduced in Theorem 4.2.1 (β) tolerates sector bounded input and output nonlinearities.

Theorem 6.2.1
Let $\xi(\cdot, \cdot), \eta(\cdot, \cdot)$ be given as in (6.6) with (unknown) $0 < \xi_1 < \xi_2, 0 < \eta_1 < \eta_2$, and let $N(\cdot): \mathbb{R} \to \mathbb{R}$ be a Nussbaum function, $p \geq 1$.

Consider the system

$$\begin{aligned}
\dot{x}(t) &= Ax(t) + b\xi(t, u(t)) &, x(0) = x_0 \\
y(t) &= cx(t),
\end{aligned} \right\} \tag{6.7}$$

with $(A, b, c) \in \mathbb{R}^{n \times n} \times \mathbb{R}^n \times \mathbb{R}^{1 \times n}$ minimum phase and $cb \neq 0$. If the adaptive feedback strategy

$$\begin{aligned}
u(t) &= -N(k(t))\eta(t, y(t)) \\
\dot{k}(t) &= |\eta(t, y(t))|^P &, k(0) = k_0
\end{aligned} \right\} \tag{6.8}$$

is applied to (6.7), for arbitrary $k_0 \in \mathbb{R}$, $x_0 \in \mathbb{R}^n$, then the closed-loop system has a solution $(x(\cdot), k(\cdot)) : [0, \omega) \to \mathbb{R}^{n+1}$ for some $\omega > 0$, and every solution has, on its maximal interval of existence $[0, \omega)$, the properties

(i) $\omega = \infty$,

(ii) $\lim_{t \to \infty} k(t) = k_\infty$ exists and is finite,

(iii) $x(\cdot) \in L_p(0, \infty) \cap L_\infty(0, \infty)$ and $\lim_{t \to \infty} x(t) = 0$.

Proof: (a):   Since the right hand side of the closed-loop system (6.7), (6.8) satisfies the Carathéodory conditions, the initial value problem possesses a solution $(x(\cdot), k(\cdot)) : [0, \omega) \to \mathbb{R}^{n+1}$, which can be maximally extended over $[0, \omega)$, for some $\omega > 0$.

(b):   For $P = 1$, an application of the inequality (2.7) yields, for some $M > 0$ and almost all $t \in [0, \omega)$,

$$\frac{1}{p}|y(t)|^P \leq M + M \int_0^t |y(s)|^P ds + cb \int_0^t |y(s)|^{P-1} \beta(y(s))\xi(s, -N(k(s))\eta(s, y(s))) \, ds.$$

We have

$$-cby(s)\xi(s, -N(k(s))\eta(s, y(s))) \leq -cb\hat{N}(k(s))\eta(s, y(s))y(s)$$

where

$$\hat{N}(k) := \begin{cases} -\xi_2 N(k) &, \text{ if } \ cbN(k) \geq 0 \\ -\xi_1 N(k) &, \text{ if } \ cbN(k) > 0. \end{cases}$$

Now

$$\eta_2^{-1}|\eta(s, y)| \leq |y| \leq \eta_1^{-1}|\eta(s, y)| \quad \text{for all} \quad s, y \in \mathbb{R} \tag{6.9}$$

yields, for $y(s) \neq 0$,

$$-cb|y(s)|^{P-2}y(s)\xi(s, -N(k(s))\eta(s, y(s))) \leq -cb\tilde{N}(k(s))|\eta(s, y(s))|^P$$

where

$$\tilde{N}(k) := \begin{cases} \hat{N}(k)\eta_2^{-(p-1)} & , \text{ if } cb\hat{N}(k) \leq 0 \\ \hat{N}(k)\eta_1^{-(p-1)} & , \text{ if } cb\hat{N}(k) < 0. \end{cases}$$

Therefore,

$$\frac{1}{p}|y(t)|^p \leq M + M\eta_1^{-p}\left[k(t) - k(0)\right] - cb\int_0^t \tilde{N}(k(s))\dot{k}(s)ds$$

for all $t \in [0,\omega)$. Since $\tilde{N}(\cdot) : \mathbb{R} \to \mathbb{R}$ is a Nussbaum function, and $\xi(\cdot,\cdot)$, $\eta(\cdot,\cdot)$ are sector bounded, the remainder of the proof follows as in the proof of Theorem 4.2.1. $\square$

In the following theorem it will be shown that the 'standard' universal adaptive stabilizer also tolerates sector bounded input and output nonlinearities, if applied to single-input, single-output, almost strictly positive real systems.

Theorem 6.2.2
Let $\xi(\cdot,\cdot)$ and $\eta(\cdot,\cdot)$ be given as in (6.6). Consider the system

$$\left. \begin{array}{rcl} \dot{x}(t) & = & Ax(t) + b\xi(t,u(t)) \quad , x(0) = x_0 \\ y(t) & = & cx(t), \end{array} \right\} \tag{6.10}$$

with $(A,b,c) \in \mathbb{R}^{n \times n} \times \mathbb{R}^n \times \mathbb{R}^{1 \times n}$ minimal and almost strictly positive real. If the adaptive feedback strategy

$$\left. \begin{array}{rcl} u(t) & = & -k(t)y(t) \\ \dot{k}(t) & = & \eta(t,y(t))^2 \quad , k(0) = k_0 \end{array} \right\} \tag{6.11}$$

is applied to (6.10), for arbitrary $k_0 \in \mathbb{R}$, $x_0 \in \mathbb{R}^n$, then the closed-loop system has a solution $(x(\cdot),k(\cdot)) : [0,\omega) \to \mathbb{R}^{n+1}$ for some $\omega > 0$, and every solution has, on its maximal interval of existence $[0,\omega)$, the properties

(i) $\omega = \infty$,

(ii) $\lim_{t \to \infty} k(t) = k_\infty$ exists and is finite,

(iii) $x(\cdot) \in L_2(0,\infty) \cap L_\infty(0,\infty)$ and $\lim_{t \to \infty} x(t) = 0$.

Proof: Existence of a maximal solution $(x(\cdot),k(\cdot)) : [0,\omega) \to \mathbb{R}^{n+1}$ follows as in the proof of Theorem 6.2.1.
By assumption there exists a $K \in \mathbb{R}$ so that

$$\begin{array}{rcl} \dot{x}(t) & = & \hat{A}x(t) + \hat{b}\left[\xi(t,u(t)) - Ky(t)\right], \\ y(t) & = & \hat{c}x(t), \end{array}$$

and $(\hat{A}, \hat{b}, \hat{c})$, defined in (3.14), is strictly positive real. Let $P = P^T \in \mathbb{R}^{n \times n}$ be the positive-definite solution of the Lur'e equations associated with $(\hat{A}, \hat{b}, \hat{c})$ for some $Q \in \mathbb{R}^{n \times m}$ and $\mu > 0$. Then the derivative of $V(x) = \langle x, Px \rangle$ along the solution component $x(t)$ of the closed-loop system

$$
\begin{aligned}
\dot{x}(t) &= \hat{A}x(t) + \hat{b}\left[\xi\left(t, -N(k(t))\eta(t, y(t))\right) - Ky(t)\right] \\
y(t) &= \hat{c}x(t)
\end{aligned}
$$

is, confer Lemma 3.1.6, for almost all $t \in [0, \omega)$,

$$
\begin{aligned}
\frac{d}{dt}V(x(t)) &= 2\langle x(t), P\hat{A}x(t)\rangle + 2\langle x(t), P\hat{b}\left[\xi\left(t, -N(k(t))\eta(t, y(t))\right) - Ky(t)\right]\rangle \\
&= -\|Q^T x(t)\|^2 - \mu V(x(t)) \\
&\quad + 2y(t) \cdot \left[\xi\left(t, -N(k(t))\eta(t, y(t))\right) - Ky(t)\right] \\
&\leq 2Ky(t)^2 - 2\tilde{N}(k(t))\eta\left(t, y(t)\right)^2 \\
&= 2Ky(t)^2 - 2\tilde{N}(k(t))\dot{k}(t),
\end{aligned}
$$

where

$$
\tilde{N}(k) := \begin{cases} \xi_1 \eta_2^{-1} N(k) & , \text{ if } N(k) \geq 0 \\ \xi_2 \eta_1^{-1} N(k) & , \text{ if } N(k) < 0. \end{cases}
$$

The remainder of the proof follows in a similar manner as part (e) of the proof of Theorem 4.2.4. $\qquad\qquad\square$

In order to use a switching decision function for stabilization in the presence of sector bounded input-output nonlinearities, the switching strategy needs a modification.

Let $1 > \lambda_1 > \lambda_2 > \ldots > 0$ be a strictly decreasing sequence with $\lim_{i \to \infty} \lambda_i = 0$, and sector bounded nonlinearities $\xi(\cdot, \cdot)$, $\eta(\cdot, \cdot)$ are given as in (6.6). For $y(\cdot) \in L_p(0, \infty)$, $p \geq 1$, and

$$
\dot{k}(t) = |\eta(t, y(t))|^p, \qquad k(0) = 0,
$$

let $\Theta(\cdot) : \mathbb{R} \to \{-1, +1\}$ be determined by the algorithm

$$
(*) \quad \left.\begin{array}{rcl}
i & := & 0 \\
\Theta(0) & := & -1 \quad , t_0 := 0 \\
t_{i+1} & := & \inf\{t > t_i | \, |\zeta(t)| \leq 1 - \lambda_i\} \\
\Theta(t) & := & \Theta(t_i) \quad \text{for all} \quad t \in [t_i, t_{i+1}) \\
\Theta(t_{i+1}) & := & -\Theta(t_i) \\
i & := & i + 1 \\
\text{go to } (*)
\end{array}\right\} \quad (6.12)
$$

where a modified switching decision function is defined by

$$\zeta(t) := \frac{\int_0^t \Theta(\tau)k(\tau)|\eta(\tau, y(\tau))|^p d\tau}{1 + \int_0^t k(\tau)|\eta(\tau, y(\tau))|^p d\tau} \in (-1, +1)$$

Theorem 6.2.3
Let $p \geq 1$, $\xi(\cdot, \cdot)$, $\eta(\cdot, \cdot)$ be given as in (6.6), and $1 > \lambda_1 > \lambda_2 > \ldots$ be a strictly decreasing sequence with $\lim_{i \to \infty} \lambda_i = 0$. Consider the system

$$\left. \begin{array}{rcl} \dot{x}(t) & = & Ax(t) + b\xi(t, u(t)) \quad , x(0) = x_0 \\ y(t) & = & cx(t), \end{array} \right\} \tag{6.13}$$

with $(A, b, c) \in \mathbb{R}^{n \times n} \times \mathbb{R}^n \times \mathbb{R}^{1 \times n}$ minimum phase and $cb \neq 0$. If the adaptive feedback strategy

$$\begin{array}{rcl} u(t) & = & -k(t)\Theta(t)\eta(t, y(t)), \\ \dot{k}(t) & = & |\eta(t, y(t))|^p \quad k(0) = 0, \end{array}$$

where $\Theta(\cdot)$ is defined via the algorithm (6.12), is applied to (6.13), for arbitrary $x_0 \in \mathbb{R}^n$, then the closed-loop system has a solution $(x(\cdot), k(\cdot)) : [0, \omega) \to \mathbb{R}^{n+1}$ for some $\omega > 0$, and every solution has, on its maximal interval of existence $[0, \omega)$, the properties

(i) $\omega = \infty$,

(ii) $\lim_{t \to \infty} k(t) = k_\infty$ exists and is finite,

(iii) $x(\cdot) \in L_p(0, \infty) \cap L_\infty(0, \infty)$ and $\lim_{t \to \infty} x(t) = 0$,

(iv) $\Theta(t)$ switches only finitely many times.

Proof: (a):   Since the right hand side of the closed-loop system satisfies the Carathéodory conditions, the initial value problem possesses a solution $(x(\cdot), k(\cdot)) : [0, \omega) \to \mathbb{R}^{n+1}$, which can be maximally extended over $[0, \omega)$, for some $\omega > 0$.

(b):   Suppose $cb > 0$, the case $cb < 0$ is proved in a similar manner. Lemma 6.1.1 and (6.9) yield, for some $M > 0$ and all $t \in [0, \omega)$,

$$\frac{1}{p}|y(t)|^p \leq M + M \int_0^t |y(s)|^p ds - cb \int_0^t k(s)\Theta(s)|y(s)|^{p-1}\beta(y(s))\eta(s, y(s))ds$$

$$\leq M + M\eta_1^{-1} \int_0^t |\eta(s, y(s))|^p ds - cb \int_0^t k(s)\bar{\Theta}(s)|\eta(s, y(s))|^p ds,$$

where

$$\bar{\Theta}(t) := \begin{cases} \eta_2^{-(p-1)}\Theta(t) & , \text{ if } \Theta(t) = 1 \\ \eta_1^{-(p-1)}\Theta(t) & , \text{ if } \Theta(t) = -1. \end{cases}$$

Set

$$\alpha := \eta_1^{-(p-1)}, \quad \text{and} \quad \beta := \eta_2^{-(p-1)}.$$

Suppose $k \notin L_\infty(0,\omega)$ and let $t_0 \in [0,\omega)$ so that $k(t_0) > 0$. The inequality above yields, for all $t \in [t_0, \omega)$,

$$\frac{1}{p}|y(t)|^p \leq M + cb\frac{\beta-\alpha}{2} + \eta_1^{-1}k(t) - cb\left[\frac{\beta-\alpha}{2} + \int_0^t \bar{\Theta}(s)k(s)\dot{k}(s)ds\right]$$

$$\leq M + k(t)\left[M\eta_1^{-1} - cb\bar{\psi}(t)\right] \tag{6.14}$$

where, for all $t \in [t_0, \omega)$,

$$\bar{\psi}(t) := \frac{\frac{1}{2}(\beta-\alpha) + \int_0^t \bar{\Theta}(\tau)k(\tau)|\eta(\tau,y(\tau))|^p d\tau}{\int_0^t |\eta(\tau,y(\tau))|^p d\tau}.$$

It is easy to see that

$$\bar{\psi}(t) = \frac{\beta-\alpha}{2}\phi(t) + \frac{\beta+\alpha}{2}\psi(t) = \frac{\beta+\alpha}{2}\left[\frac{\beta-\alpha}{\beta+\alpha} + \frac{\psi(t)}{\phi(t)}\right]\phi(t) \tag{6.15}$$

where, for all $t \in [t_0, \omega)$,

$$\psi(t) = \frac{\int_0^t \Theta(\tau)k(\tau)\dot{k}(\tau)d\tau}{\int_0^t \dot{k}(\tau)d\tau}, \quad \phi(t) = \frac{1 + \int_0^t k(\tau)\dot{k}(\tau)d\tau}{\int_0^t \dot{k}(\tau)d\tau}.$$

Now by construction

$$\limsup_{k\to\omega} \zeta(t) = \limsup_{k\to\omega} \frac{\psi(t)}{\phi(t)} = 1,$$

and hence, by (6.15),

$$\liminf_{k\to\omega} \bar{\psi}(t) = +\infty.$$

This contradicts (6.14) and it follows that $k(\cdot) \in L_\infty(0,\omega)$. Therefore, $\omega = \infty$, and the remainder of the proof follows in a similar manner as the proof of Lemma 2.1.8. $\square$

Almost strictly positive real systems can even be subjected to *multivariable* sector bounded input nonlinearities and the standard universal adaptive stabilizer is still applicable. This will be proved in the remainder of this section.

Let $\xi(\cdot) : \mathbb{R}^m \rightarrow \mathbb{R}^m$ be a continuous map satisfying the sector boundedness condition

$$\langle \xi(u) - \Delta_1 u, \xi(u) - \Delta_2 u \rangle \leq 0 \quad \text{for all} \quad u \in \mathbb{R}^m \tag{6.16}$$

where

$$\Delta_i = \text{diag}\left\{\delta_1^i, \ldots, \delta_m^i\right\}, \quad \delta_1^i, \ldots, \delta_m^i > 0 \quad \text{for} \quad i = 1, 2.$$

This is a multivariable extension of the scalar sector bounded condition in (6.6). It is easy to see that

$$\|\xi(u)\|^2 + \alpha_1 \|u\|^2 \leq \alpha_2 \langle u, \xi(u) \rangle \quad \text{for all} \quad u \in \mathbb{R}^m \tag{6.17}$$

where

$$\alpha_1 := \min_{1 \leq j \leq m}\left\{\delta_j^1 \delta_j^2\right\}, \qquad \alpha_2 := \max_{1 \leq j \leq m}\left\{\delta_j^1 + \delta_j^2\right\}.$$

Theorem 6.2.4
Let $\xi(\cdot) : \mathbb{R}^m \rightarrow \mathbb{R}^m$ be a sector bounded continuous function satisfying (6.16) and consider the system

$$\left. \begin{aligned} \dot{x}(t) &= Ax(t) + B\xi(u(t)) \quad , x(0) = x_0 \\ y(t) &= Cx(t), \end{aligned} \right\} \tag{6.18}$$

with $(A, B, C) \in \mathbb{R}^{n \times n} \times \mathbb{R}^{n \times m} \times \mathbb{R}^{m \times n}$ minimal, almost strictly positive real, and $\text{rk}B = m \leq n$. If the adaptive feedback mechanism

$$\left. \begin{aligned} u(t) &= -k(t)y(t), \\ \dot{k}(t) &= \|y(t)\|^2 \quad , k(0) = k_0 \end{aligned} \right\} \tag{6.19}$$

is applied to (6.18), then, for arbitrary $k_0 \in \mathbb{R}$, $x_0 \in \mathbb{R}^n$, the closed-loop system has a solution $(x(\cdot), k(\cdot)) : [0, \omega) \rightarrow \mathbb{R}^{n+1}$, for some $\omega > 0$, which satisfies on its maximal interval of existence $[0, \omega)$,

(i) $\omega = \infty$,

(ii) $\lim_{t \to \infty} k(t) = k_\infty$ exists and is finite,

(iii) $x(\cdot) \in L_2(0, \infty) \cap L_\infty(0, \infty)$ and $\lim_{t \to \infty} x(t) = 0$.

Proof: (a)   Since the right hand side of the closed-loop system (6.18), (6.19) satisfies the Carathéodory conditions, the initial value problem has a solution $(x(\cdot), k(\cdot)) : [0, \omega) \rightarrow \mathbb{R}^{n+1}$, which can be maximally extended over $[0, \omega)$, for some $\omega > 0$.

(b): Since the system is almost strictly positive real there exists a $K \in \mathbb{R}^{m \times m}$, confer (3.13), so that the solution of the closed-loop system satisfies

$$\begin{aligned}
\dot{x}(t) &= \hat{A}x(t) + \hat{B}\left[\xi\left(-k(t)y(t)\right) - Ky(t)\right] \quad , x(0) = x_0 \\
y(t) &= \hat{C}x(t),
\end{aligned}$$

with strictly positive real $(\hat{A}, \hat{B}, \hat{C})$ defined in (3.14). Let $P = P^T \in \mathbb{R}^{n \times n}$ be a positive-definite solution of the Lur'e equations associated with $(\hat{A}, \hat{B}, \hat{C})$, for some $Q \in \mathbb{R}^{n \times m}$ and $\mu > 0$. The derivative of $V(x) = \langle x, Px \rangle$ along the closed-loop system satisfies, for almost all $t \in [0, \omega)$,

$$\begin{aligned}
\frac{d}{dt}V(x(t)) &= 2\langle x(t), P\hat{A}x(t)\rangle + 2\left\langle x(t), P\hat{B}\left[\xi\left(-k(t)y(t)\right) - Ky(t)\right]\right\rangle \\
&\leq -\|Q^T x(t)\|^2 - \mu V(x(t)) + 2\langle y(t), \xi\left(-k(t)y(t)\right)\rangle + 2\|K\|\|y(t)\|^2.
\end{aligned}$$

Suppose that $k(\cdot) \notin L_\infty(0, \omega)$. Let $t^* \in [0, \omega)$ so that $k(t^*) > 0$. Then the inequality (6.17) yields

$$\langle y(t), \xi\left(-k(t)y(t)\right)\rangle \leq -(\alpha_2 k(t))^{-1}\|\xi\left(-k(t)y(t)\right)\|^2 - \frac{\alpha_1}{\alpha_2}k(t)\|y(t)\|^2,$$

and thus we have, for almost all $t \in [0, \omega)$,

$$\frac{d}{dt}V(x(t)) \leq -2\left[\frac{\alpha_1}{\alpha_2}k(t) - \|K\|\right]\dot{k}(t),$$

whence by integration

$$V(x(t)) \leq V(x(t^*)) - 2\int_{t^*}^{t}\left[\frac{\alpha_1}{\alpha_2}k(s) - \|K\|\right]\dot{k}(s)ds$$

and hence, for all $t \in [t^*, \omega)$ with $k(t) > k(t^*)$,

$$V(x(t)) \leq V(x(t^*)) + [k(t) - k(t^*)]\left[2\|K\| - \frac{\alpha_1}{\alpha_2}(k(t) + k(t^*))\right].$$

Since $k(\cdot)$ is assumed to be unbounded on $[0, \omega)$, the above inequality yields a contradiction, and hence $k(\cdot) \in L_\infty(0, \omega)$. The remainder of the proof follows as in part (e) of the proof of Theorem 4.2.4. $\square$

6.3 λ-Tracking controller

In this section, we shall show that the type of λ-tracking controller introduced in Section 5.2, and simple modifications of it, is capable of tolerating large classes of nonlinearities, including output corrupted noise belonging to $\mathcal{W}^{1,\infty}(\mathbb{R}, \mathbb{R}^m)$. The results are also interesting, if only λ-stabilization is considered. At the expense of λ-stabilization, instead of asymptotic stabilization, a much larger class of nonlinearities is tolerated than considered in Section 6.1.

Before we will prove these results, we try to make it plausible why even the robustness and well posedness results obtained in Section 6.1 cannot be expected for the internal model based asymptotic tracking controller considered in Section 5.1. The reason is simply, that a converging gain would mean that the terminal system is a nonlinear system, and this cannot be expected to track a signal generated by a linear system. However, it can be shown that the asymptotic tracking controller tolerates signals $d(\cdot) \in L_p(0, \infty)$ in the nominal system: The series interconnection between (6.1), for $g_1(\cdot, \cdot) \equiv 0$, $g_2(\cdot, \cdot) \equiv 0$, $h(\cdot, \cdot) \equiv 0$, and the internal model (5.2) is given by

$$\begin{aligned} \dot{\bar{x}}(t) &= \bar{A}\bar{x}(t) + \bar{B}v(t) + \bar{d}(t) &, \bar{x}(0) = (x_0^T, \xi_0^T)^T \\ y(t) &= \bar{C}\bar{x}(t) + \bar{D}v(t) \end{aligned}$$

with $\bar{A}, \bar{B}, \bar{C}, \bar{D}, \bar{x}$ as in (5.3) and $\bar{d} = \left(d^T, 0\right)^T$.

The proof of the universal adaptive tracking controller in Theorem 5.1.3 is mainly based on a transformation of the tracking problem into a stabilization problem. The same transformation is possible in the present nonlinear setup and instead of (5.10) we obtain

$$\begin{aligned} \dot{x}_e(t) &= \bar{A}x_e(t) + \bar{B}v(t) + \bar{d}(t) \\ e(t) &= \bar{C}x_e(t) + \bar{D}v(t) - y_{\text{ref}}^-(t). \end{aligned}$$

Now asymptotic tracking can be proved as in the proof of Theorem 5.1.3 by using the inequality in Lemma 6.1.1, which takes into account the signal $\bar{d}(\cdot)$.

In the remainder of this section, we will show that the λ-tracking controller, presented in Section 5.2, tolerates a fairly rich class of nonlinearities.

The systems to be considered are of the form

$$\left.\begin{aligned} \dot{x}(t) &= Ax(t) + g(t, x(t)) + B\left[\xi\left(t, u(t)\right) + f(t, x(t))\right], \quad x(0) = x_0 \\ y(t) &= Cx(t) + n(t) \end{aligned}\right\} \quad (6.20)$$

with $(A, B, C) \in \mathbb{R}^{n \times n} \times \mathbb{R}^{n \times m} \times \mathbb{R}^{m \times n}$ minimum phase. See Figure 6.3.

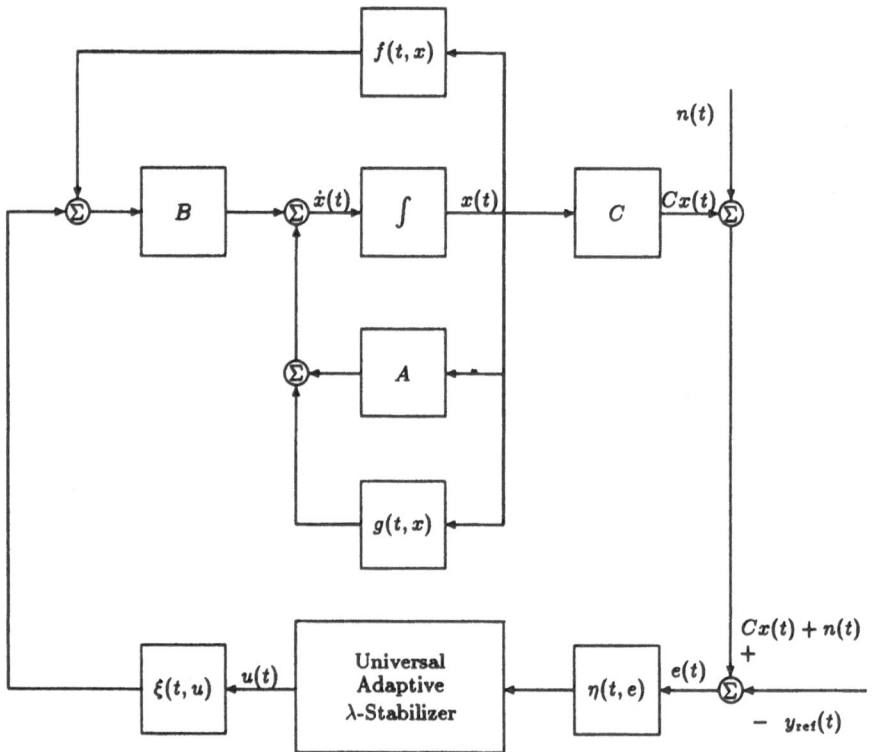

Figure 6.3: Adaptive λ-tracking for nonlinear perturbed systems in the presence of noise

As in Section 5.2, the reference signal $y_{\text{ref}}(\cdot)$ to be λ-tracked is assumed to belong to $\mathcal{W}^{1,\infty}(\mathbb{R}, \mathbb{R}^m)$, no bound respectively bound on its derivative need to be known explicitly.

Throughout this section, the nonlinearities

$$
\left.
\begin{array}{llll}
g(\cdot, \cdot) & : & \mathbb{R} \times \mathbb{R}^n \to \mathbb{R}^n, & \|g(t, x)\| \leq \hat{g}[1 + \|Cx\|] \\
f(\cdot, \cdot) & : & \mathbb{R} \times \mathbb{R}^n \to \mathbb{R}^m, & \|f(t, x)\| \leq \hat{f}[\|x\| + \bar{f}(t, Cx)]
\end{array}
\right\}
$$

$$(6.21)$$

are assumed to be Carathéodory functions which, for some $\hat{g}, \hat{f} \geq 0$, unknown to the designer, satisfy the boundedness conditions for almost all $t \in \mathbb{R}$ and for all $x \in \mathbb{R}^n$.

$\bar{f}(\cdot, \cdot), : \mathbb{R} \times \mathbb{R}^m \to \mathbb{R}^m$ is an (unknown) Carathéodory function which satisfies for some *known* continuous function $\tilde{f} : \mathbb{R}^m \to [0, \infty)$:

$$
\left.
\begin{array}{c}
\text{for every } R > 0, \text{ there is an unknown scalar } \alpha_R \text{ such that,} \\
\bar{f}(t, u + v) \leq \alpha_R \tilde{f}(u) \quad \text{for all } v, u \in \mathbb{R}^m \text{ with } \|v\| \leq R \\
\text{and for almost all } t \in \mathbb{R}.
\end{array}
\right\}
$$

$$(6.22)$$

The function $\tilde{f}(\cdot)$ will be used in the feedback strategy to compensate the nonlinearity $f(\cdot, \cdot)$. $f(\cdot, \cdot)$ may be, for example, a polynomial of degree not exceeding r, say, on the components of $Cx(t)$ with t-dependent coefficients of L_∞ class. Then we would choose $\tilde{f}(v) = 1 + \|v\|^r$. If $f(\cdot, \cdot) \equiv 0$, then we choose $\tilde{f}(\cdot) \equiv 0$.

Note that the nonlinearity $g(\cdot, \cdot)$ covers input corrupted noise belonging to $\mathcal{W}^{1,\infty}(\mathbb{R}, \mathbb{R}^m)$.

If single-input, single-output systems are considered, the controller also tolerates sector bounded nonlinearities $\xi(\cdot, \cdot), \eta(\cdot, \cdot)$ as defined in (6.6), the sector bounds need not to be known.

The control strategy will be of the form

$$e(t) = y(t) - y_{\text{ref}}(t)$$

$$u(t) = -N(k(t)) \left[\eta(t, e(t)) + \tilde{f}\left(\eta(t, e(t)) + y_{\text{ref}}(t)\right) s_{\tilde{\lambda}}(e(t))e(t) \right]$$

$$\dot{k}(t) = D_\Lambda(e(t)) \left[\|\eta(t, e(t))\| + \tilde{f}\left(\eta(t, e(t)) + y_{\text{ref}}(t)\right) \right] \qquad , k(0) = k_0$$

with $D_\Lambda(e)$ defined in (5.18), or a simplification, depending on whether the system class is multivariable, what is known about the spectrum of CB, and which nonlinearities are considered.

$N(\cdot) : \mathbb{R} \to \mathbb{R}$ is a Nussbaum function, possibly scaling-invariant.

For $\lambda > 0$, let

$$s_\lambda(\cdot) : \mathbb{R}^m \to [0, \lambda^{-1}]$$

be any continuous function with the properties

$$\left.\begin{array}{rcl} s_\lambda(e) = 0 & \Longleftrightarrow & e = 0 \\ d_\lambda(e) \geq 0 & \Longrightarrow & s_\lambda(e) = \|e\|^{-1}, \end{array}\right\} \tag{6.23}$$

with $d_\lambda(\cdot)$ defined in (5.12).
A simple example of one such function is

$$s_\lambda(e) = \left\{ \begin{array}{ll} \|e\|^{-1} & , \text{if } \|e\| \geq \lambda \\ \lambda^{-1} & , \text{if } \|e\| < \lambda. \end{array} \right.$$

Applying the coordinate transformation of Lemma 2.1.3 to (6.20) yields, for

$$e(t) = Cx(t) + n(t) - y_{\text{ref}}(t),$$

$$\left.\begin{array}{l} \dot{e}(t) = A_1 e(t) + A_2 z(t) + g_1(t, x(t)) \\ \qquad + CB\left[\xi\left(t, u(t)\right) + f(t, x(t))\right] \ , e(0) = Cx_0 + n(0) - y_{\text{ref}}(0) \\ \dot{z}(t) = A_3 e(t) + A_4 z(t) + g_2(t, x(t)) \ , z(0) = N x_0 \end{array}\right\} \tag{6.24}$$

where

$$\begin{array}{rcl} g_1(t, x(t)) & := & Cg(t, x(t)) + \dot{n}(t) - \dot{y}_{\text{ref}}(t) - A_1\left[n(t) - y_{\text{ref}}(t)\right] \\ g_2(t, x(t)) & := & Ng(t, x(t)) - A_3\left[n(t) - y_{\text{ref}}(t)\right]. \end{array}$$

Since

$$S^{-1}x(t) = \left((Cx(t))^T, z(t)^T\right)^T = \left((e(t) - n(t) + y_{\text{ref}}(t))^T, z(t)^T\right)^T,$$

the second argument of $g_1 t, \cdot)$ and $g_2(t, \cdot)$ can be expressed by the new variables $e(t)$, $z(t)$. For notational convenience we leave the argument $x(t)$ in $g_1(t, \cdot)$ and $g_2(t, \cdot)$.

By assumption on the noise and reference signal, there exists $M_1 \geq 0$ so that

$$\|n(\cdot) - y_{\text{ref}}(\cdot)\|_{L_\infty(0,\infty)} + \|\dot{n}(\cdot) - \dot{y}_{\text{ref}}(\cdot)\|_{L_\infty(0,\infty)} \leq M_1$$

and hence, by (6.21) and

$$Cx(t) = e(t) + y_{\text{ref}}(t) - n(t),$$

we have

$$\|g_i(t, x(t))\| \leq M_2 + M_2\|e(t)\| \quad \text{for} \quad i = 1, 2 \tag{6.25}$$

where

$$M_2 := [\|C\| + \|N\|]\hat{g}(1 + M_1) + M_1[1 + \|A_1\|].$$

The proofs of the main results of this section require the following technical lemma.

Lemma 6.3.1
Let $\rho > 0$, $P = P^T \in \mathbb{R}^{m \times m}$ be positive-definite and

$$V_\rho(e) := \frac{1}{2} D_\rho(e)^2 = \begin{cases} \frac{1}{2} \left(\|e\|_P - \rho \right)^2 & , \text{ if } \|e\|_P \geq \rho \\ 0 & , \text{ if } \|e\|_P < \rho. \end{cases}$$

If the initial value problem (6.20) has an almost continuous solution $x(\cdot) : [0, \omega) \to \mathbb{R}^n$ for some $\omega > 0$, then the derivative of $V_\rho(e)$ along the solution component $e(t)$ of (6.24) satisfies, for almost all $t \in [0, \omega)$,

$$\frac{d}{dt} V_\rho(e(t)) \leq M D_\rho(e(t)) \left[\|e(t)\| + \|z(t)\| + \tilde{f}(e(t) + y_{\text{ref}}(t)) \right]$$
$$+ D_\rho(e(t)) \|e(t)\|_P^{-1} \langle e(t), PCB\xi(t, u(t)) \rangle$$

for some $M > 0$.

Proof: For notational convenience we introduce

$$\theta_\rho(t) := D_\rho(e(t)) \|e(t)\|_P^{-1} e(t).$$

Using (6.25), the bound on $\bar{f}(\cdot)$, and

$$\|\theta_\rho(t)\| \leq \rho^{-1} \|P^{1/2}\| \|e(t)\| \|\theta_\rho(t)\|,$$

we conclude, for almost all $t \in [0, \omega)$,

$$\begin{aligned}
\frac{d}{dt} V_\rho(e(t)) &= \langle \theta_\rho(t), P\dot{e}(t) \rangle \\
&\leq \|\theta_\rho(t)\| \left[\|PA_1\| \|e(t)\| + \|PA_2\| \|z(t)\| \right] \\
&\quad + \|\theta_\rho(t)\| \left[M_2 \|P\| \rho^{-1} \|P^{1/2}\| \|e(t)\| + M_2 \|P\| \|e(t)\| \right] \\
&\quad + \|\theta_\rho(t)\| \|PCB\| \hat{f} \left[\|S\| \left(\|e(t)\| + M_1 \rho^{-1} \|P^{1/2}\| \|e(t)\| + \|z(t)\| \right) \right] \\
&\quad + \|\theta_\rho(t)\| \|PCB\| \hat{f} \alpha_{M_1} \tilde{f}(e(t) + y_{\text{ref}}(t)) \\
&\quad + \langle \theta_\rho(t), PCB\xi(t, u(t)) \rangle \\
&\leq \|\theta_\rho(t)\| M_3 \left[\|z(t)\| + \|e(t)\| + \tilde{f}(e(t) + y_{\text{ref}}(t)) \right] \\
&\quad + \langle \theta_\rho(t), PCB\xi(t, u(t)) \rangle
\end{aligned}$$

with

$$\begin{aligned}
M_3 &:= \|PA_1\| + \|PA_2\| + M_2 \|P\| \rho^{-1} \|P^{1/2}\| + M_2 \|P\| \\
&\quad + \|PCB\| \hat{f} \left[2\|S\| + M_1 \rho^{-1} \|P^{1/2}\| + \alpha_{M_1} \right].
\end{aligned}$$

Now the lemma follows for $M = M_3 \mu_{\min}(P)^{-1/2}$.   □

In the following theorem it will be shown that a simple modification of the λ-tracking controller (5.26) is also applicable if the nominal system is subjected to noise corrupted output, and nonlinearities of the form (6.21) in the input and state.

Theorem 6.3.2
Consider the nonlinear perturbed system

$$\left.\begin{array}{rcl} \dot{x}(t) & = & Ax(t) + g(t, x(t)) + B\left[u(t) + f(t, x(t))\right] \quad , x(0) = x_0 \\ y(t) & = & Cx(t) + n(t) \end{array}\right\} \quad (6.26)$$

with $(A, B, C) \in \mathbb{R}^{n \times n} \times \mathbb{R}^{n \times m} \times \mathbb{R}^{m \times n}$ minimum phase. Suppose there exist positive-definite $P = P^T$, $Q = Q^T \in \mathbb{R}^{m \times m}$ and $\beta \in \{-1, +1\}$ so that

$$PCB + (CB)^T P = 2\beta Q.$$

Let $g(\cdot, \cdot)$ and $f(\cdot, \cdot)$ satisfy (6.21) and (6.22) for a known function $\tilde{f} : \mathbb{R}^m \to [0, \infty)$. If $p, q > 0$ are known so that

$$p\|e\| \le \|e\|_P \le q\|e\| \quad \text{for all} \quad e \in \mathbb{R}^m,$$

and $N(\cdot) : \mathbb{R} \to \mathbb{R}$ is an arbitrary scaling-invariant Nussbaum function, $\lambda > 0$, $s_{p\lambda q-1}(\cdot); [0, p^{-1}\lambda^{-1}q]$ satisfying (6.23), then the adaptive feedback mechanism

$$\left.\begin{array}{l} e(t) = y(t) - y_{\text{ref}}(t), \\ u(t) = -N(k(t))\left[1 + \tilde{f}\left(e(t) + y_{\text{ref}}(t)\right) s_{p\lambda q-1}(e(t))\right] e(t) \\ \dot{k}(t) = D_{p\lambda}(e(t))\left[\|e(t)\| + \tilde{f}\left(e(t) + y_{\text{ref}}(t)\right)\right] \qquad , k(0) = k_0 \end{array}\right\}$$
$$(6.27)$$

applied to (6.26), for arbitrary $x_0 \in \mathbb{R}^n$, $k_0 \in \mathbb{R}$, $n(\cdot)$, $y_{\text{ref}}(\cdot) \in W^{1,\infty}(\mathbb{R}, \mathbb{R}^m)$, yields a solution $(x(\cdot), k(\cdot)) : [0, \omega) \to \mathbb{R}^{n+1}$ of the closed-loop system, for some $\omega > 0$, and every solution has on its maximal interval of existence $[0, \omega)$ the properties

(i) $\omega = \infty$,

(ii) $\lim_{t \to \infty} k(t) = k_\infty$ exists and is finite,

(iii) $x(\cdot), k(\cdot) \in L_\infty(0, \infty)$,

(iv) the error $e(t)$ approaches the closed ball $\overline{B}_\lambda(0)$ as $t \to \infty$.

Note that the only difference between the feedback mechanism (6.27) and (5.26) is that $\tilde{f}(\cdot, \cdot)$ is inserted in order to cope with the nonlinearity $f(\cdot, \cdot)$. If $f(\cdot, \cdot) \equiv 0$, then (5.26) can be applied.

Note also that $e(t) = Cx(t) + n(t) - y_{\text{ref}}(t)$ is controlled towards the ball $\bar{B}_\lambda(0)$. Therefore, if an upper bound M on the noise $n(\cdot)$ is known a priori, λ should not be chosen smaller than M since this would mean the 'true output' $Cx(t)$ is tracking the noise corrupted reference signal.

Proof of Theorem 6.3.2:

(a): Since the right hand side of the closed-loop system (6.26), (6.27) satisfies the Carathéodory conditions, there exists a solution $(x(\cdot), k(\cdot)) : [0, \omega) \to \mathbb{R}^{n+1}$, and every solution can be extended to its maximal interval of existence $[0, \omega)$, for some $\omega > 0$.

(b): Since

$$D_{p\lambda}(e) > 0 \quad \Longrightarrow \quad s_{p\lambda q^{-1}}(e) = \|e\|^{-1},$$

we have

$$D_{p\lambda}(e(t))\|e(t)\|_P^{-1}\langle e(t), PCBu(t)\rangle$$
$$= -\beta N(k(t))D_{p\lambda}(e(t))\|e(t)\|_P^{-1}\|e(t)\|_Q^2 \left[1 + \tilde{f}(e(t) + y_{\text{ref}}(t))\|e(t)\|^{-1}\right]$$
$$\leq -\tilde{N}(k(t))D_{p\lambda}(e(t)) \left[\|e(t)\| + \tilde{f}(e(t) + y_{\text{ref}}(t))\right]$$
$$\leq -\tilde{N}(k(t))\dot{k}(t),$$

where

$$\tilde{N}(k) := \begin{cases} \beta N(k)\|Q\|p^{-1} & \text{, if } -\beta N(k) \geq 0 \\ \beta N(k)\mu_{\min}(Q)q^{-1} & \text{, if } -\beta N(k) < 0. \end{cases}$$

Arguing as in part (b) of the proof of Theorem 5.2.4, there exists $M_1 > 0$ such that

$$\int_0^t D_{p\lambda}(e(s))\|z(s)\|ds \leq M_1 \int_0^t D_{p\lambda}(e(s))\|e(s)\|ds.$$

Integration of the inequality in Lemma 6.3.1 for $\xi(t, u(t)) = u(t)$, and inserting the above inequalities, yields

$$V_{p\lambda}(e(t)) \leq V_{p\lambda}(e(0)) - \int_0^t \tilde{N}(k(s))\dot{k}(s)ds$$

$$+ M(M_1 + 1) \int_0^t D_{p\lambda}(e(s)) \left[\|e(s)\| + \tilde{f}(e(s) + y_{\text{ref}}(s))\right] ds$$

$$\leq M_2 + M_2[k(t) - k(0)] - \int_{k(0)}^{k(t)} \tilde{N}(k(\tau))d\tau, \tag{6.28}$$

where

$$M_2 := V_{p\lambda}(e(0)) + M(M_1 + 1).$$

(c):   If $k(\cdot) \notin L_\infty(0, \omega)$, then, for $k(t) > k(0)$, (6.28) yields

$$V_{p\lambda}(e(t)) \leq M_2 + [k(t) - k(0)] \left[M_2 - \frac{1}{k(t) - k(0)} \int_{k(0)}^{k(t)} \tilde{N}(\tau) d\tau \right],$$

and since $\tilde{N}(\cdot)$ is a Nussbaum function, the right hand side becomes negative, thus contradicting the non-negativeness of the left hand side. This proves $k(\cdot) \in L_\infty(0, \omega)$.

(d):   Since $k(\cdot) \in L_\infty(0, \omega)$ it follows that $e(\cdot) \in L_\infty(0, \omega)$ and hence, by (6.25), $g_2(t, x(t))$ is essentially bounded in $[0, \omega)$. Thus it follows from the second equation in (6.24) and exponential stability of A_4, that $z(\cdot) \in L_\infty(0, \omega)$. Now we may conclude the assertions (i)-(iii).

(e):   It remain to prove (iv). Since

$$D_{p\lambda}(e) \leq D_{p\lambda}(e)p\lambda^{-1}\|e\|_P \leq D_{p\lambda}(e)p\lambda^{-1}q\|e\|$$

and

$$D_{p\lambda}(e)\|e\|_P^{-1}\|e\| \leq D_{p\lambda}(e)p^{-1},$$

by boundedness of $(k(\cdot), x(\cdot))$, and Lemma 6.3.1, there exists a $M_3 > 0$ such that, for almost all $t \geq 0$,

$$\frac{d}{dt}V_{p\lambda}(e(t)) \leq -p^{-1}D_{p\lambda}(e)\|e\| + M_3 D_{p\lambda}(e)\|e\| \leq -p^{-1}D_{p\lambda}(e)\|e\| + M_3\dot{k}(t).$$

Differentiation of the sign-indefinite function

$$W(\cdot, \cdot) : \mathbb{R}^{m+1} \to \mathbb{R}, \qquad (e, k) \mapsto V_{p\lambda}(e) - M_3 k,$$

yields, by Lemma 5.2.3 (i), for almost all $t \geq 0$,

$$\frac{d}{dt}W(e(t), k(t)) \leq -d_\lambda(e(t))\|e(t)\| \leq 0 \qquad (6.29)$$

and so $W(e(\cdot), k(\cdot))$ is monotone.
At this stage, we could apply LaSalle's Invariance Principle for non-autonomous systems, see LaSalle (1976), similar to part (e) of the proof of Theorem 5.2.2. However, we like to proceed with an explicit argument, tailored to the case in hand.
By boundedness of $(e(\cdot), z(\cdot), k(\cdot))$, (6.25), the bound on $f(\cdot, \cdot)$, and the first equation in (6.24) there exists $R > 0$ such that

$$\|\dot{e}(t)\| \leq R \quad \text{for almost all} \quad t \geq 0.$$

Moreover, the *bounded* solution $t \mapsto (e(t), z(t), k(t))$ must tend, as $t \to \infty$, to its non-empty ω-limit set Ω. Hence we will prove assertion (iv) (that is, $d_\lambda(e(t)) \to 0$ as $t \to \infty$) by showing that Ω is contained in the set $\Sigma =$

$\{(e, z, k) \in \mathbb{R}^{n+1} | \|e\| \le \lambda\}.$

Seeking a contradiction, suppose $\Omega \not\subset \Sigma$. Then there exists $(\bar{e}, \bar{z}, \bar{k}) \in \Omega$ and $\varepsilon > 0$ such that $d_\lambda(\bar{e})\|\bar{e}\| > 2\varepsilon$. By continuity, there exists $\delta > 0$ such that

$$\|\xi - \bar{e}\| < \delta \quad \Longrightarrow \quad d_\lambda(\xi)\|\xi\| > \varepsilon.$$

Since $(\bar{e}, \bar{z}, \bar{k})$ is an ω-limit point, there exists a sequence $\{t_j\}_{j \in \mathbb{N}}$ with $t_j \to \infty$ and

$$\lim_{j \to \infty} (e(t_j), z(t_j), k(t_j)) \to (\bar{e}, \bar{z}, \bar{k}).$$

By continuity of $W(\cdot, \cdot)$,

$$W(e(t_j), k(t_j)) - W(\bar{e}, \bar{k}) < \frac{\varepsilon\delta}{4R} \tag{6.30}$$

for all j sufficiently large. Let j^* be such that $\|e(t_j) - \bar{e}\| < \frac{1}{2}\delta$ for all $j > j^*$. Observe that, for all $j > j^*$ and all $t \ge t_j$,

$$
\begin{aligned}
\|e(t_j) - \bar{e}\| &\le \|e(t) - e(t_j)\| + \|e(t_j) - \bar{e}\| \\
&\le \int_{t_j}^{t} \|\dot{e}(s)\| ds + \|e(t_j) - \bar{e}\| \\
&\le R|t - t_j| + \frac{1}{2}\delta.
\end{aligned}
$$

Therefore, for all $j > j^*$,

$$t \in [t_j, t_j + (\delta/3R)] \quad \Longrightarrow \quad \|e(t) - \bar{e}\| < \delta \quad \Longrightarrow \quad d_\lambda(e(t))\|e(t)\| > \varepsilon.$$

By (6.29), we now have, for all $j > j^*$,

$$W(e(t_j), k(t_j)) - W(\bar{e}, \bar{k}) \ge \int_{t_j}^{t_j + (\delta/3R)} d_\lambda(e(t))\|e(t)\| dt \ge \frac{\varepsilon\delta}{3R}$$

which contradicts (6.30). Therefore, $\Omega \subset \Sigma$ and so $d_\lambda(e(t)) \to 0$ as $t \to \infty$. This completes the proof. $\qquad \square$

As we have seen in Theorem 5.2.4 for unperturbed systems, the control strategy in Theorem 6.3.2 can be simplified if it is known that (A, B, C) satisfies $\sigma(CB) \subset \mathbb{C}_+$. This is also true for a nonlinearly perturbed system (6.20) in the presence of corrupted noise.

Theorem 6.3.3
Consider the nonlinearly perturbed system (6.26) and suppose that $(A, B, C) \in \mathbb{R}^{n \times n} \times \mathbb{R}^{n \times m} \times \mathbb{R}^{m \times n}$ is minimum phase, $\sigma(CB) \subset \mathbb{C}_+$, $f(\cdot, \cdot)$, $g(\cdot, \cdot)$ satisfy (6.21) and (6.22) for some known continuous function $\hat{f} : \mathbb{R}^m \to [0, \infty)$. If

$\lambda > 0$, and $s_\lambda(\cdot)$ satisfies (6.23), then the adaptive feedback mechanism (6.27) can be replaced by the simpler version

$$\left.\begin{aligned}
e(t) &= y(t) - y_{\text{ref}}(t), \\
u(t) &= -k(t)\left[1 + \tilde{f}\left(e(t) + y_{\text{ref}}(t)\right) s_\lambda(e(t))\right] e(t) \\
\dot{k}(t) &= d_\lambda(e(t))\left[\|e(t)\| + \tilde{f}\left(e(t) + y_{\text{ref}}(t)\right)\right] \qquad , k(0) = k_0
\end{aligned}\right\} \quad (6.31)$$

and applied to (6.26) the same conclusions as in Theorem 6.3.2 are valid.

Proof: (a):   Since the right hand side of the closed-loop system (6.26), (6.31) satisfies the Carathéodory conditions, there exists a solution $(x(\cdot), k(\cdot))$: $[0, \omega) \to \mathbb{R}^{n+1}$, and every solution can be extended to its maximal interval of existence $[0, \omega)$, for some $\omega > 0$.

(b):   We shall show boundedness of $k(\cdot)$. Seeking a contradiction, suppose $k(\cdot) \notin L_\infty(0, \omega)$. Then, by monotonicity, there exists a $t_0 \in [0, \omega)$ such that $k(t) > 0$ for all $t \in [t_0, \omega)$. Lett $P = P^T \in \mathbb{R}^{m \times m}$ be the positive-definite solution of

$$PCB + (CB)^T P = 2I_m,$$

and set

$$p := \mu_{\min}(P^{1/2}), \qquad q := \mu_{\max}(P^{1/2}).$$

Now Lemma 6.3.1 yields, for almost all $t \in [t_0, \omega)$ and some $M > 0$,

$$\begin{aligned}
\frac{d}{dt}V_{p\lambda}(e(t)) \leq{}& MD_{p\lambda}(e(t))\left[\|e(t)\| + \|z(t)\| + \tilde{f}\left(e(t) + y_{\text{ref}}(t)\right)\right] \\
&- q^{-1}k(t)D_{p\lambda}(e(t))\|e(t)\|\left[1 + \tilde{f}\left(e(t) + y_{\text{ref}}(t)\right) s_\lambda(e(t))\right].
\end{aligned}$$

Since

$$D_{p\lambda}(e) \geq 0 \quad \Longrightarrow \quad pq^{-1}\lambda \leq \|e\|,$$

we have, for almost all $t \in [t_0, \omega)$,

$$\begin{aligned}
\frac{d}{dt}V_{p\lambda}(e(t)) \leq{}& MD_{p\lambda}(e(t))\left[\|e(t)\| + \|z(t)\| + \tilde{f}\left(e(t) + y_{\text{ref}}(t)\right)\right] \\
&- q^{-1}\gamma k(t)D_{p\lambda}(e(t))\left[\|e(t)\| + \tilde{f}\left(e(t) + y_{\text{ref}}(t)\right)\right],
\end{aligned}$$

where

$$\gamma := \min\left\{s_\lambda(e(t))\|e(t)\| \mid pq^{-1}\lambda \leq \|e(t)\| \leq \lambda \text{ and } t \in [t_0, \omega)\right\} \in (0, 1].$$

In part (b) of the proof of Theorem 5.2.4 we have shown that

$$\int_{t_0}^{t} D_{p\lambda}(e(s))\|z(s)\| ds \leq M_3 \int_{t_0}^{t} D_{p\lambda}(e(s))\|e(s)\| ds$$

for some $M_3 > 0$ and all $t \in t_0, \omega)$. Therefore, integration of $\frac{d}{ds}V_{p\lambda}(e(s))$ over $[t_0, t)$ for $t \in t_0, \omega)$ yields

$$V_{p\lambda}(e(t)) \leq V_{p\lambda}(e(t_0))$$
$$+ \int_{t_0}^{t} [M(M_3 + 1) - q^{-1}\gamma k(s)]D_{p\lambda}(e(s)) \left[\|e(s)\| + \tilde{f}(e(s) + y_{\mathrm{ref}}(s))\right] ds.$$

Choose $t_1 \in [t_0, \omega)$ such that

$$M(M_3 + 1) - q^{-1}\gamma k(s) < 0 \quad \text{for all} \quad s \in [t_1, \omega).$$

By Lemma 5.2.3 (i) we have, for all $t \in [t_1, \omega)$

$$V_{p\lambda}(e(t)) \leq M_4 + p \int_{t_1}^{t} [M(M_3 + 1) - q^{-1}\gamma k(s)]\dot{k}(s)ds$$

$$= M_4 + pM(M_3 + 1)[k(t) - k(t_1)] - pq^{-1}\gamma\frac{1}{2}[k(t)^2 - k(t_1)^2],$$

where

$$M_4 := V_{p\lambda}(e(t_0))$$
$$+ \int_{t_0}^{t_1} [M(M_3 + 1) - q^{-1}\gamma k(s)]D_{p\lambda}(e(s)) \left[\|e(s)\| + \tilde{f}(e(s) + y_{\mathrm{ref}}(s))\right] ds.$$

Since $k(\cdot) \notin L_\infty(0, \omega)$, the right hand side of the above inequality takes negative values, thus contradicting the non-negativeness of the left hand side. This proves boundedness of $k(\cdot)$ on $[0, \omega)$.

(c): As in part (d) of the proof of Theorem 6.3.2, the assertions (i)-(iii) follow.

(d): Since $e(\cdot)$, $y_{\mathrm{ref}}(\cdot)$, and $k(\cdot)$ are essentially bounded, by Lemma 6.3.1 there exists $M_5 > 0$ such that

$$\frac{d}{dt}V_{q\lambda}(e(t)) \leq M_5 D_{q\lambda}(e(t))\|e(t)\|$$
$$\leq M_5 q d_\lambda(e(t))\|e(t)\|$$
$$\leq M_5 q\dot{k}(t),$$

where we have used Lemma 5.2.3 (ii). Defining

$$W(e, k) = V_{q\lambda}(e) - [M_5 q + 1]k,$$

the remainder follows in a similar way as the proof of part (e) of Theorem 6.3.2. This completes the proof. □

In the remainder of this section it will be shown that, for single-input, single-output nonlinearly perturbed systems in the presence of output corrupted noise, the λ-tracking controller also tolerates sector bounded output nonlinearities, and, if $f(\cdot, \cdot) \equiv 0$, sector bounded input and output nonlinearities .

Theorem 6.3.4
Let $\xi(\cdot, \cdot)$, $\eta(\cdot, \cdot)$ be Carathéodory function as in (6.6) with (unknown) $0 < \xi_1 < \xi_2$, $0 < \eta_1 < \eta_2$. Suppose the system

$$\left. \begin{array}{rcl} \dot{x}(t) & = & Ax(t) + b\xi\,(t, u(t)) \quad , x(0) = x_0 \\ y(t) & = & cx(t) + n(t), \end{array} \right\} \tag{6.32}$$

with $(A, b, c) \in \mathbb{R}^{n \times n} \times \mathbb{R}^n \times \mathbb{R}^{1 \times n}$, is minimum phase and $cb > 0$. If $\lambda > 0$, and the adaptive feedback mechanism

$$\left. \begin{array}{rcl} e(t) & = & y(t) - y_{\text{ref}}(t) \\ u(t) & = & -k(t)\eta\,(t, e(t)) \\ \dot{k}(t) & = & d_\lambda\,(\eta(t, e(t)))\,|\eta(t, e(t))| \quad , k(0) = k_0 \end{array} \right\} \tag{6.33}$$

is applied to (6.32), for arbitrary $x_0 \in \mathbb{R}^n$, $k_0 \in \mathbb{R}$, $n(\cdot)$, $y_{\text{ref}}(\cdot) \in \mathcal{W}^{1,\infty}(\mathbb{R}, \mathbb{R})$, then there exists a solution $(x(\cdot), k(\cdot)) : [0, \omega) \to \mathbb{R}^{n+1}$ of the closed-loop system for some $\omega > 0$, and every solution has on its maximal interval of existence $[0, \omega)$ the properties

(i) $\omega = \infty$,

(ii) $\lim_{t \to \infty} k(t) = k_\infty$ exists and is finite,

(iii) $x(\cdot), k(\cdot) \in L_\infty(0, \infty)$,

(iv) the error $e(t)$ approaches the interval $[-\lambda, +\lambda]$ as $t \to \infty$.

Proof: (a):   Since the right hand side of the closed-loop system (6.32), (6.33) satisfies the Carathéodory conditions, there exists a solution $(x(\cdot), k(\cdot))$ and every solution can be extended to its maximal interval of existence $[0, \omega)$, for some $\omega > 0$.

(b):   Suppose $k(\cdot) \notin L_\infty(0, \omega)$. Then there exists $t_0 \in [0, \omega)$ so that $k(t_0) \geq 0$. By Lemma 6.3.1 and the sector bounds on $\eta(t, e(t))$ and $\xi(t, u(t))$, we have, for $P = 1$ and for almost all $t \in [t_0, \omega)$ and some $M > 0$,

$$\begin{aligned} \frac{d}{dt} V_\rho(e(t)) & \leq & M d_\rho(e(t))[|e(t)| + \|z(t)\|] \\ & & + cb\, d_\rho(e(t))|e(t)|^{-1} e(t)\xi\,(t, -k(t)\eta(t, e(t))) \\ & \leq & M d_\rho(e(t))[|e(t)| + \|z(t)\|] - cb\,\xi_1\eta_1 k(t) d_\rho(e(t))|e(t)|. \end{aligned} \tag{6.34}$$

Arguing as in part (b) of the proof of Theorem 5.2.2 and using Lemma 5.2.1 yields, for some $M_2 > 0$ and all $t \in [t_0, \omega)$,

$$\int_{t_0}^{t} d_\rho(e(s))\|z(s)\|ds \leq M_2 \int_{t_0}^{t} d_\rho(e(s))|e(s)|ds,$$

and hence, by integration of $\frac{d}{ds} V_\rho(e(s))$,

$$V_\rho(e(t)) \leq V_\rho(e(t_0)) + \int_{t_0}^{t} [M(1 + M_2) - cb\,\xi_1\eta_1 k(s)]\, d_\rho(e(s))|e(s)|ds.$$

Note that
$$-d_{\eta_2^{-1}\lambda}(e) \leq -\eta_2^{-1} d_\lambda(\eta(t,e)) \quad \text{for all} \quad e \in \mathbb{R}.$$

Since we have assumed $k(\cdot) \notin L_\infty(0, \omega)$, there exists $t_1 \in [t_0, \omega)$ such that

$$M(1 + M_2) - cb\,\xi_1\eta_1 k(s) \leq 0 \quad \text{for all} \quad s \in [t_1, \omega),$$

and hence, for $\rho := \xi_1^{-1}\lambda$ and all $t \in [t_1, \omega)$ so that $k(t) > k(t_1)$,

$$
\begin{aligned}
V_{\xi_1^{-1}\lambda}(e(t)) \;\leq\;& V_{\xi_1^{-1}\lambda}(e(t_1)) \\
&+ \int_{t_1}^{t} [M(1 + M_2) - cb\,\xi_1\eta_1 k(s)]\, \eta_2^{-1} d_\lambda(\eta(s, e(s)))\eta_2^{-1}|\eta(s, e(s)|ds \\
=\;& V_{\xi_1^{-1}\lambda}(e(t_1)) \\
&+ \eta_2^{-2}[k(t) - k(t_1)]\left[M(1 + M_2) - \frac{cb\,\xi_1\eta_1}{k(t) - k(t_1)} \int_{k(t_1)}^{k(t)} \tau d\tau \right].
\end{aligned}
$$

Now unboundedness of $k(\cdot)$ yields that the right hand side of the above inequality takes negative values, thus contradicting non-negativeness of the left hand side. This proves $k(\cdot) \in L_\infty(0, \omega)$.

(c): Since $\xi(t, u)$ is sector bounded from above, assertions (i)-(iii) follow in a similar manner as part (d) of the proof of Theorem 6.3.2.

(d): By (6.34), we have, for almost all $t \geq 0$ and some $M_3 > 0$,

$$\frac{d}{dt} V_{\eta_1^{-1}\lambda}(e(t)) \leq M_3 d_{\eta_1^{-1}\lambda}(e(t))|e(t)| \leq M_3 \eta_1^{-2} d_\lambda(\eta(t, e(t)))|\eta(t, e(t))| = M_3 \dot{k}(t).$$

Defining

$$W(\cdot, \cdot) : \mathbb{R}^{m+1} \to \mathbb{R}, \qquad (e, k) \mapsto V_{\eta_1^{-1}\lambda}(e) - [M_3 + 1]k,$$

the remainder of the proof follows similar to part (d) of the proof of Theorem 6.3.2. This completes the proof. $\qquad\qquad \square$

At the expense of allowing only sector bounded *input* nonlinearities, a nonlinearity $f(t, x)$ is tolerated and the sign of the high-frequency gain need not to be known.

Theorem 6.3.5
Let $\xi(\cdot, \cdot)$ be Carathéodory function as in (6.6) with (unknown) $0 < \xi_1 < \xi_2$, $N(\cdot) : \mathbb{R} \to \mathbb{R}$ a scaling-invariant Nussbaum function, and let $\bar{f}(\cdot, \cdot)$ satisfy (6.22) for some known continuous function $\tilde{f} : \mathbb{R}^m \to [0, \infty)$. If $\lambda > 0$, $s_\lambda(\cdot)$ satisfies (6.33), and the adaptive feedback mechanism

$$
\left.
\begin{aligned}
e(t) &= y(t) - y_{\text{ref}}(t), \\
u(t) &= -N(k(t)) \left[1 + \tilde{f}\left(e(t) + y_{\text{ref}}(t)\right) s_\lambda(e(t)) \right] e(t) \\
\dot{k}(t) &= d_\lambda(e(t)) \left[|e(t)| + \tilde{f}\left(e(t) + y_{\text{ref}}(t)\right) \right] \qquad , k(0) = k_0
\end{aligned}
\right\}
\tag{6.35}
$$

is applied to the system

$$
\left.
\begin{aligned}
\dot{x}(t) &= Ax(t) + b\left[\xi\left(t, u(t)\right) + f(t, x(t)\right] \quad , x(0) = x_0 \\
y(t) &= cx(t) + n(t),
\end{aligned}
\right\}
\tag{6.36}
$$

with $(A, b, c) \in \mathbb{R}^{n \times n} \times \mathbb{R}^n \times \mathbb{R}^{1 \times n}$ minimum phase and $cb \neq 0$, for arbitrary $x_0 \in \mathbb{R}^n$, $k_0 \in \mathbb{R}$, $n(\cdot)$, $y_{\text{ref}}(\cdot) \in \mathcal{W}^{1,\infty}(\mathbb{R}, \mathbb{R})$, then there exists a solution $(x(\cdot), k(\cdot)) : [0, \omega) \to \mathbb{R}^{n+1}$ of the closed-loop system for some $\omega > 0$, and every solution has on its maximal interval of existence $[0, \omega)$ the properties

(i) $\omega = \infty$,

(ii) $\lim_{t \to \infty} k(t) = k_\infty$ exists and is finite,

(iii) $x(\cdot), k(\cdot) \in L_\infty(0, \infty)$,

(iv) the error $e(t)$ approaches the interval $[-\lambda, +\lambda]$ as $t \to \infty$.

Proof: (a):   Since the right hand side of the closed-loop system (6.35), (6.36) satisfies the Carathéodory conditions, there exists a solution $(x(\cdot), k(\cdot))$ and every solution can be extended to its maximal interval of existence $[0, \omega)$, for some $\omega > 0$.

(b):   By Lemma 6.3.1 and the sector bounds on $\xi(t, u(t))$ we have, for $P = 1$, for almost all $t \in [0, \omega)$ and some $M > 0$,

$$
\frac{d}{dt} V_\lambda(e(t)) \leq M d_\lambda(e(t)) \left[|e(t)| + \|z(t)\| + \tilde{f}\left(e(t) + y_{\text{ref}}(t)\right) \right]
$$
$$
+ cb\, d_\lambda(e(t)) \frac{e(t)}{|e(t)|} \xi\left(t, -N(k(t)) \left[1 + \tilde{f}\left(e(t) + y_{\text{ref}}(t)\right) s_\lambda(e(t)) \right] e(t) \right).
$$

For $|e| \geq \lambda$ we have $s_\lambda(e)|e| = 1$ and thus

$$
cb\, d_\lambda(e) \frac{e}{|e|} \xi\left(t, -N(k) \left[1 + \tilde{f}\left(e + y_{\text{ref}}\right) s_\lambda(e) \right] e \right) \leq -\tilde{N}(k) d_\lambda(e) \left[|e| + \tilde{f}\left(e + y_{\text{ref}}\right) \right]
$$

where

$$\tilde{N}(k) := \begin{cases} \xi_1 cbN(k) & , \text{if } cbN(k) \geq 0 \\ \xi_2 cbN(k) & , \text{if } cbN(k) < 0. \end{cases}$$

By using the same argument as in part (b) of the proof of Theorem 5.2.2

$$\int_0^t d_\lambda(e(s))\|z(s)\|ds \leq M_1 \int_0^t d_\lambda(e(s))|e(s)|ds,$$

for some $M_1 > 0$. Suppose $k(\cdot) \notin L_\infty(0, \omega)$. Integration of $\frac{d}{ds}V_\lambda(e(s))$ over $[0, t]$ yields, for all $t \in [0, \omega)$ so that $k(t) > k(0)$,

$$\begin{aligned} V_\lambda(e(t)) &\leq V_\lambda(e(0)) + M(M_1 + 1)\int_0^t d_\lambda(e(s))\left[|e(s)| + \tilde{f}(e(t) + y_{\text{ref}}(t))\right] \\ &\quad - \int_0^t d_\lambda(e(s))\tilde{N}(k(s))\left[|e(s)| + \tilde{f}(e(s) + y_{\text{ref}}(s))\right] ds \\ &\leq V_\lambda(e(0)) + [k(t) - k(0)]\left[M(M_1 + 1) - \frac{1}{k(t) - k(0)}\int_{k(0)}^{k(t)} \tilde{N}(\tau)d\tau\right]. \end{aligned}$$

(c): The assertions (i)-(iii) follow in a similar way as in part (b)-(d) of the proof of Theorem 6.3.2.

(d): By essential boundedness of $e(\cdot)$, $z(\cdot)$, and $k(\cdot)$ and the inequalities in (b), we have, for some $M_2 > 0$ and almost all $t \geq 0$,

$$\frac{d}{dt}V_\lambda(e(t)) \leq M_2 d_\lambda(e(t))|e(t)| \leq M_2\dot{k}(t).$$

Now the remainder of the proof follows, by using

$$W(e, k) := V_\lambda(e) - (M_2 + 1)k,$$

as in part (e) of the proof of Theorem 6.3.2.
This completes the proof. □

6.4   Notes and References

Section 6.1:   Lemma 6.1.1 is due to Ilchmann and Owens (1992). Well posedness and robustness has been considered, apart for almost strictly positive real systems and for the nonlinearity $h(t, u(t))$, by: Helmke and Prätzel-Wolters (1988) for scalar systems; Owens et al. (1987), Prätzel-Wolters et al. (1989), Ilchmann and Owens (1991a, 1992) for various system classes and adaptive feedback mechanisms. For adaptive stabilization of scalar nonlinear minimum phase systems see Mårtensson (1990) and Nikitin and Schmid (1990), and for certain multivariable, nonlinear, minimum phase systems see Saberi and Lin (1990), Khalil and Saberi (1987).

Section 6.2:   Adaptive stabilization in the presence of sector bounded nonlinearities has been studied for single-input, single-output, retarded systems by Logemann and Owens (1988) and by Logemann (1990). Theorem 6.2.1 is from Ilchmann and Owens (1992). The switching strategy (6.12) and the proof of Theorem 6.2.3 is a simplification of a result due to Ilchmann and Owens (1991a).

Section 6.3:   That the universal adaptive tracking controller using an internal model tolerates signals $d(\cdot) \in L_p(0, \infty)$ in the state equation of the system has been shown by Helmke et al. (1990) for single-input, single-output systems.

All results, apart those considering sector bounded input and output nonlinearities, are due to Ilchmann and Ryan (1992). Ryan (1992a) has proved, by using set-valued maps and differential inclusions, that the λ-tracking controller tolerates more general input nonlinearities, encompassing hysteresis effects. For asymptotic tracking in the presence of nonlinearities, but using discontinuous output feedback, see Ryan (1992).

Chapter 7

Performance

The purpose of the present chapter is twofold: we shall illustrate the dynamics of the universal adaptive stabilizers introduced in Chapter 4 so that a better qualitative understanding is achieved and, furthermore, we shall derive theoretical results to improve various aspects of the performance considerably. The purpose is not, to give a complete comparison of all different adaptation mechanisms.

In Section 7.1, we show that the increase of certain parameters in the adaptation law (either p in $k(t) = \int_0^t \|y(s)\|^p ds$ or ε in $k(t) = \varepsilon \|y(t)\|^2 + \int_0^t \|y(s)\|^2 ds$) leads to a considerably reduced maximum value of $\|y(t)\|$ and $k(t)$, at the expense that the system needs longer to settle down. The 'ε-modification' is also achieved for asymptotic and λ-tracking controllers.

In Section 7.2, we modify, at the expense of derivative feedback, the feedback strategy so that the output is less than an arbitrary, small prespecified positive constant after an arbitrary, prespecified period of time with an arbitrary, short upper bound of the overshoot.

All simulations have been performed on an IBM AT 386-compatible computer and the operating system MS-DOS 5.0 with the software: SIMULAB/SIMULINK and MATLAB for Windows, Version 1.1/1.2.1, The MathWorks Inc. Massachusetts ©91/92.

7.1 Improved output and gain behaviour

In this section, we first discuss the effect of $p \geq 1$ in $\dot{k}(t) = \|y(t)\|^p$ when switching strategies introduced in Section 4.2 are applied. Furthermore, we prove that the gain adaptation can be modified to $k(t) = \varepsilon \|y(t)\|^2 +$

$\int_0^t \|y(s)\|^2 ds + \eta(0)$, having the benefit that already for 'small' ε the transient behaviour of the gain and the output is considerably improved. This result is also valid, in a similar manner, for λ-tracking and asymptotic tracking.

Throughout this chapter, the main example considered we will be the strictly proper system

$$
\dot{x}(t) = \begin{bmatrix} 10 & 10.5 & 10 \\ 0 & -0.5 & -10 \\ 0 & 10 & -0.5 \end{bmatrix} x(t) + \begin{bmatrix} 1 \\ 0 \\ 0 \end{bmatrix} u(t) \quad , x(0) = \begin{pmatrix} 0.1 \\ 0.1 \\ 0.1 \end{pmatrix} \quad (7.1)
$$

$$
y(t) = \begin{bmatrix} 1, & 1, & 0 \end{bmatrix} x(t).
$$

This system is non-controllable and non-observable, but minimum phase and therefore stabilizable and detectable. The matrix in the south-east corner of A in $\dot{x} = Ax = bu$, $y = cx$, that is

$$
A_4 = \begin{bmatrix} -0.5 & -10 \\ 10 & -0.5 \end{bmatrix},
$$

has eigenvalues $-0.5 \pm 10\,i$. Under the feedback $u(t) = -ky(t)$, the closed-loop system has the spectrum

$$
\sigma(A_k) = \{-0.5 + 10i, -0.5 - 10i, 10 - k\}, \quad \text{where} \quad A_k := A - kbc.
$$

Therefore, every $k > 10$ is a stabilizing gain.

Example 7.1.1
For the simple adaptive controller

$$
u(t) = -k(t)y(t), \qquad \dot{k}(t) = y(t)^2, \qquad k(0) = k_0 = 0 \qquad (7.2)
$$

applied to (7.1), the transient behaviour is shown in Figure 7.1. In the beginning, $y(t)$ blows up since, for small t, and therefore small $k(t)$, the time-varying linear part of the closed-loop system

$$
\dot{x}(t) = [A - k(t)bc]x(t) \qquad (7.3)
$$

is unstable, this increases the gain $k(t)$ until (7.3) gets stable, $x(t)$ decays exponentially and $k(t)$ converges to a finite limit. It is typical that $k(t)$ becomes larger than is necessary for stabilization. This is due to the integral $\int_{t'}^\infty y(s)^2 ds$ when $y(t)$ starts decaying exponentially at t', $k(t')$ a stabilizing gain.

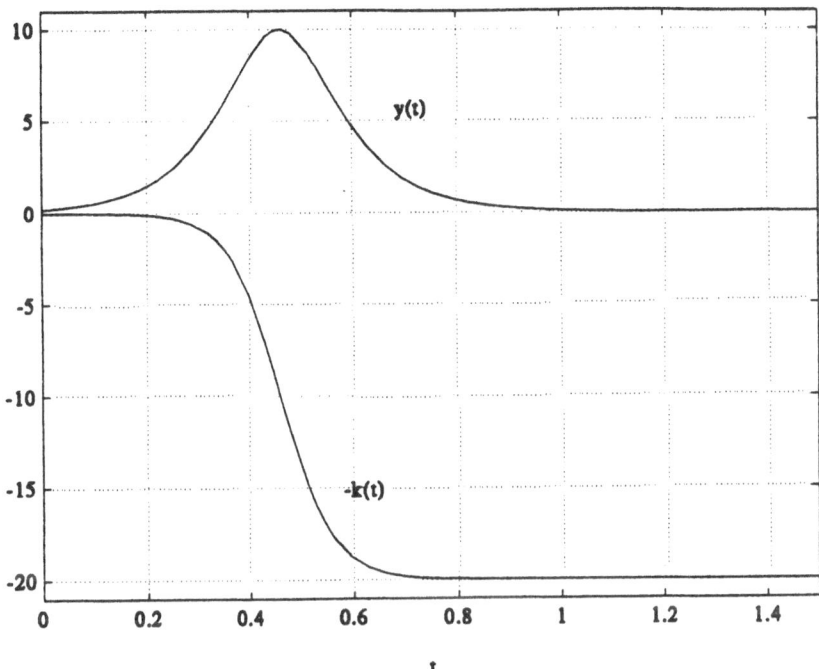

Figure 7.1:Output and gain evolution for $u(t) = -k(t)y(t)$, $\dot{k}(t) = y(t)^2$

Example 7.1.2
In Theorem 4.2.1, adaptive stabilization was proved for the adaptation law
$\dot{k}(t) = \|y(t)\|^p$, for arbitrary $p \geq 1$. The effect of large p is that, if the system
is unstable and $\|y(t)\| > 1$, then $k(t)$ will increase rapidly. But as soon as a
stablizing $k(t')$ is reached, $\|y(t)\|$ decreases rapidly and when $\|y(t)\| < 1$, it
pays a very little contribution to the integral $\int_{t'}^{\infty} y(s)^p ds$. Thus k_∞ is expected
to be smaller if p is large. These features are illustrated in Figure 7.2 where the
transient behaviour of $y(t)$ and $k(t)$ is plotted for the closed-loop system (7.1),

$$u(t) = -k(t)y(t), \quad \dot{k}(t) = |y(t)|^p, \quad k_0 = 0,$$

and $p = 1, 1.5, \ldots, 5$. For $p = 1$, the maximum value of $y(t)$ is 60, whereas it is
considerably smaller for larger p, even for $p = 1.5$, due to the faster dynamics
of the system. In all cases, the limiting gain is $k_\infty \approx 20$ (double as large as
necessary for stabilization), but for p large this final value is reached earlier.

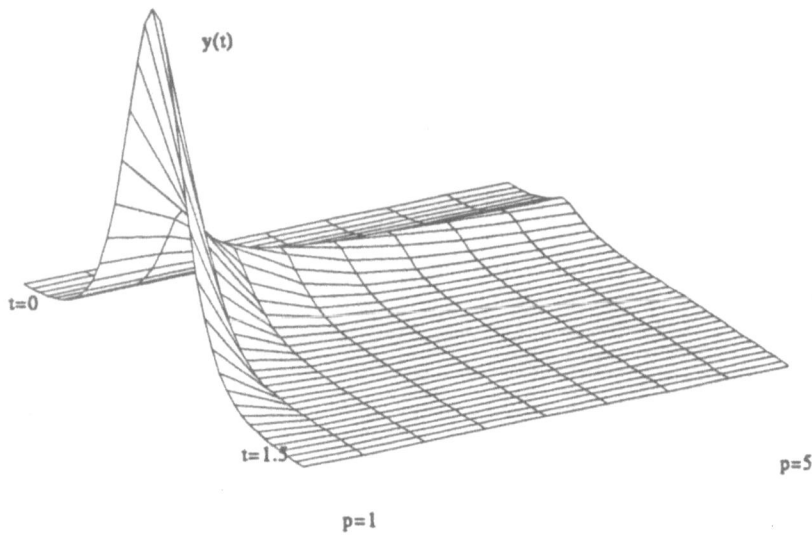

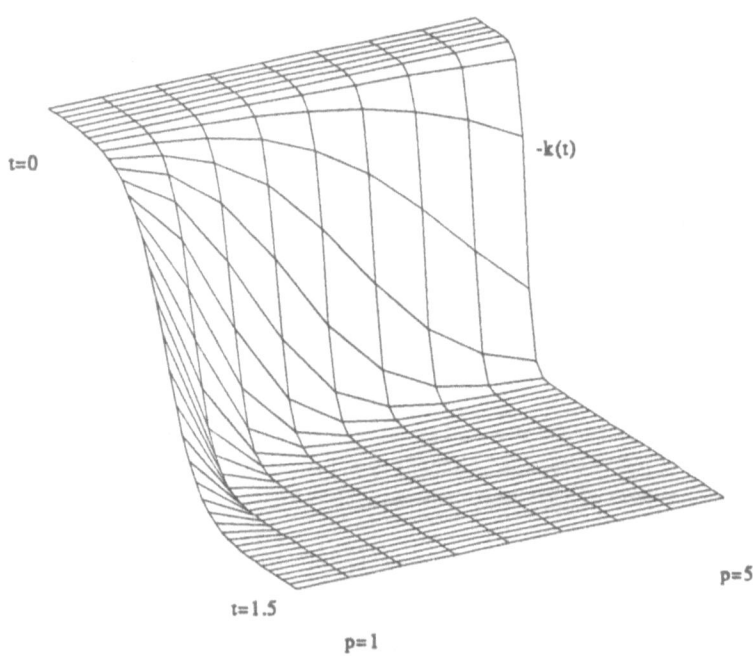

Figure 7.2: Stabilization via $u(t) = -k(t)y(t)$, $\dot{k}(t) = |y(t)|^p$ for $p = 1, 1.5, \ldots, 5$

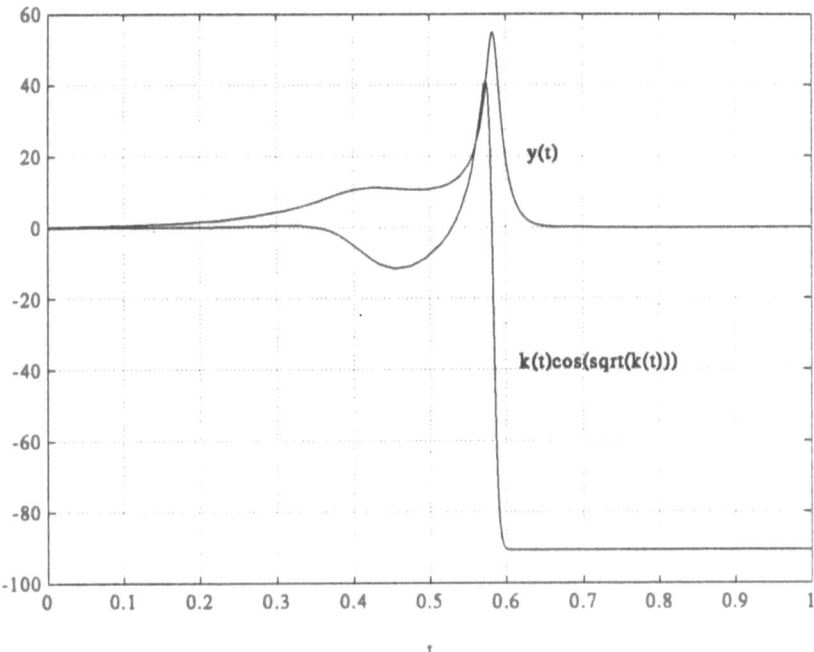

Figure 7.3: Output and gain evolution for $u(t) = -N(k(t))y(t), \dot{k}(t) = y(t)^2$

Example 7.1.3

For sake of completeness, we also assume that *sign cb* is unknown and consider stabilization of (7.1) by a Nussbaum-type feedback strategy

$$
\begin{aligned}
u(t) &= -N(k)y(t) &, N(k) &= k(t)\cos\sqrt{k(t)} \\
\dot{k}(t) &= \|y(t)\|^p &, k_0 &= 0,
\end{aligned}
$$

for $p = 2$. In Figure 7.3, the gain and output evolution is shown. For t small, the sign of $\cos\sqrt{k(t)}$ has a destabilizing effect, $y(t)$ blows up, so that at $t \approx 0.35$, $\cos\sqrt{k(t)}$ changes sign, now the feedback has a stabilizing effect but there is not enough time and the gain is not large enough so that $y(t)$ can settle down. It switches at $t \approx 0.53$ again. This time $y(t)$ is large, so that $\int_0^t y(s)^2 ds \cdot \cos\sqrt{k(t)}$ is strongly destabilizing, thus the sign switches at $t \approx 0.58$ and from then on the gain is so negative, that $y(t)$ is very quickly forced to zero so that no more switching occurs. Note that, compared to Figure 7.1, where the high-frequency gain was assumed to be known, the terminal gain $k_\infty = \lim_{t\to\infty} k(t)$ is much larger than is necessary for stabilization. This is due to the destabilizing effect when the sign of the Nussbaum function in 'wrong'.

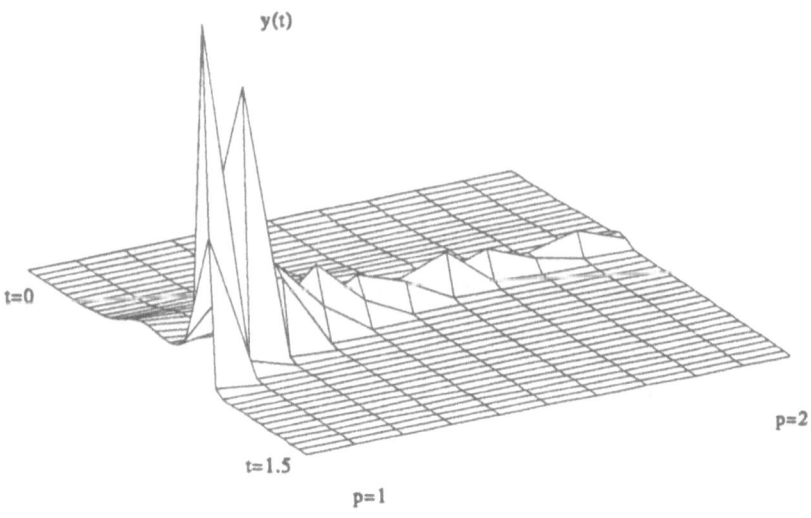

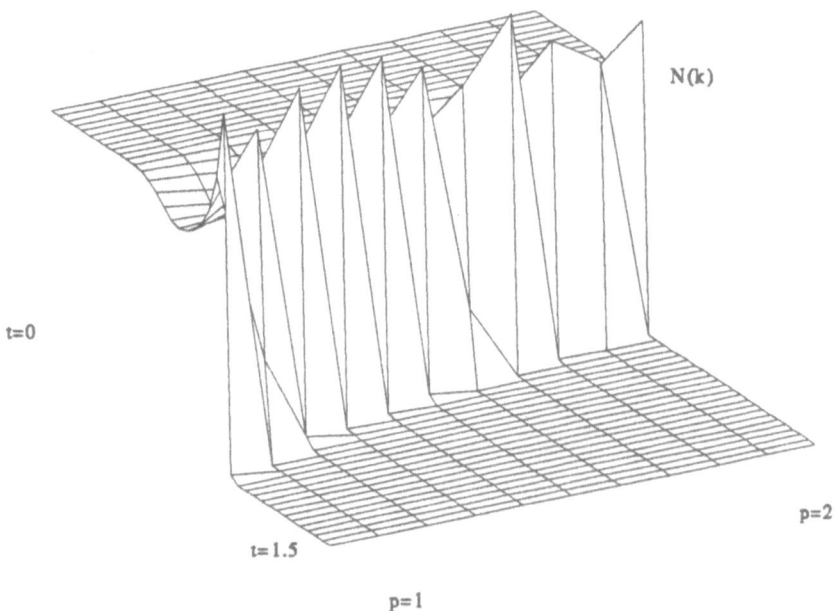

Figure 7.4: Stabilization via $u(t) = -N(k(t))y(t)$, $\dot{k}(t) = |y(t)|^p$ and $p = 1, 1.1, \ldots, 2$

Example 7.1.4
We consider the same feedback strategy as in Example 7.1.3 but vary $p = 1, 1.5, \ldots, 5$ in the gain adaptation. As in the known sign case, the transient behaviour becomes better when p is increased. See Figure 7.4. The terminal gain is still very large and settles down at $N(k_\infty) \approx -80$. Due to the switching, which causes instability if the sign is 'wrong', the peak for $y(t)$ in case of $p = 1$ is reached at $y(0.55) \approx 1500$. For larger p the behaviour of $y(t)$ is considerably improved, but only slightly for $N(k)$.

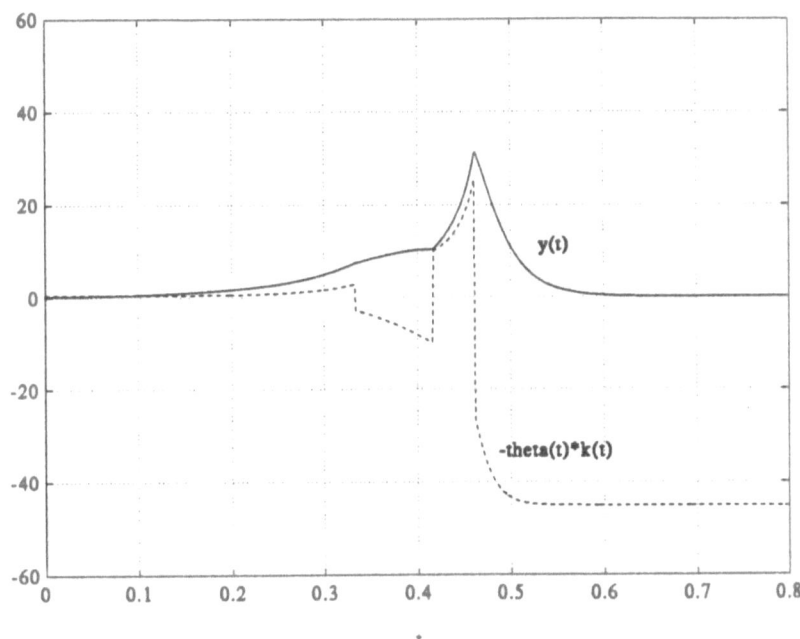

Figure 7.5: Stabilization via switching decision function and
$$u(t) = -\Theta(t)\, k(t) y(t)$$

Example 7.1.5
We also consider stabilization of (7.1) via the switching strategy based on the switching decision function introduced in Section 4.3. If the sequence of

thresholds is given by $\lambda_i = 2, 8, 16, \ldots$ and $k_0 = 0.5$, then the feedback strategy

$$
\begin{aligned}
u(t) &= -k(t)\Theta(t)y(t) \\
\dot{k}(t) &= y(t)^2, \qquad\qquad k(0) = k_0
\end{aligned}
$$

where $\Theta(t)$ is defined by the switching function $\psi(t)$ and the switching algorithm (4.25), produces the gain and output evolution as shown in Figure 7.5. For $\psi(t)$ see Figure 4.2. For this particular example, the behaviour is very similar to that for continuous feedback shown in Figure 7.3.

In the following theorem we will show that the adaptation law considered in Theorem 4.2.1 can be modified by changing $k(t) = \int_0^t \|y(s)\|^2 ds + k_0$ to $k(t) = \varepsilon\|y(t)\|^2 + \int_0^t \|y(s)\|^2 ds + \eta_0$ for arbitrary $\varepsilon > 0$. See Figure 7.6. The effect of this modification, which will be illustrated and discussed for several examples after the proof of Theorem 7.1.6, can be summarized as follows: $k(t)$ is no longer monotone, instead $k(t)$ increases much faster if the system is unstable, due to the immediate contribution which $\varepsilon\|y(t)\|^2$ pays. Thus we would expect a smaller overshoot of $\|y(t)\|$.

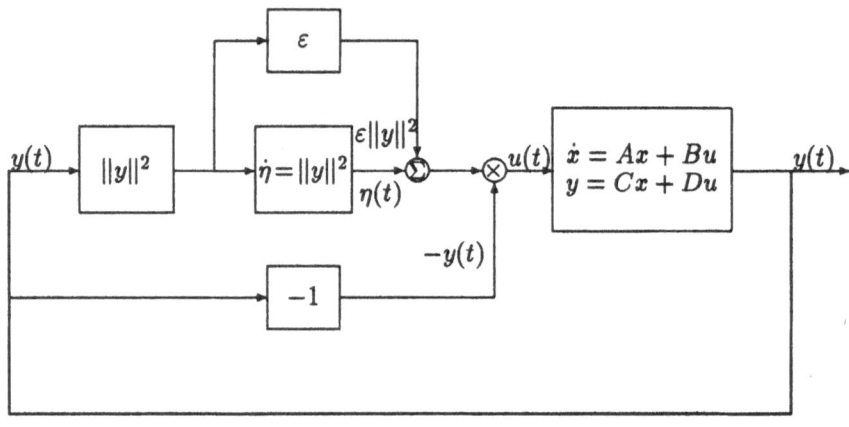

Figure 7.6: Universal Adaptive Stabilizer with ε-modification

Theorem 7.1.6
Suppose the system

$$
\left.
\begin{aligned}
\dot{x}(t) &= Ax(t) + Bu(t) \qquad , x(0) = x_0 \\
y(t) &= Cx(t) + Du(t),
\end{aligned}
\right\}
\tag{7.4}
$$

with $(A, B, C, D) \in \mathbb{R}^{n\times n} \times \mathbb{R}^{n\times m} \times \mathbb{R}^{m\times n} \times \mathbb{R}^{m\times m}$, satisfies one of the conditions

(I)   it is minimum phase, $D = 0$, and $\sigma(CB) \subset \mathbb{C}_+$,

(II)   it is minimal, almost strictly positive real, $\mathrm{rk}\, B = m \leq n$, and $D + D^T \geq 0$,

(III) it is minimal, minimum phase, $\mathrm{rk}\, B = m \leq n$, and

        a) there exists an orthogonal matrix $[Z, W]$, $Z \in \mathbb{R}^{m \times (m-r)}, W \in \mathbb{R}^{m \times r}$, $r = \mathrm{rk}\, D$ such that

$$[Z, W]^T D [Z, W] = \begin{bmatrix} 0, & 0 \\ 0, & W^T D W \end{bmatrix}$$

        b) $\det(Z^T C B Z) \neq 0$,

        c) $W^T [D + D^T] W \geq 0$.

If the feedback strategy

$$
\begin{aligned}
u(t) &= -k(t) y(t) \\
k(t) &= \varepsilon \| y(t) \|^2 + \int_0^t \| y(s) \|^2 ds + \eta_0,
\end{aligned}
$$

for arbitrary $\varepsilon, \eta_0 \geq 0$, $x_0 \in \mathbb{R}^n$, is applied to (7.4), then the closed-loop system has the properties

(i) The unique solution $(x(\cdot), k(\cdot)) : [0, \infty) \to \mathbb{R}^{n+1}$ exists,

(ii) $\lim_{t \to \infty} k(t) = k_\infty$ exists and is finite,

(iii) $x(\cdot) \in L_2(0, \infty) \cap L_\infty(0, \infty)$ and $\lim_{t \to \infty} x(t) = 0$.

Proof: (a):   Suppose (I) is valid. In order to show existence of a solution we rewrite the closed-loop system as a differential equation. Differentiation of $k(t)$ yields

$$\dot{k}(t) = 2\varepsilon \langle y(t), \dot{y}(t) \rangle + \| y(t) \|^2, \quad k(0) = \eta_0 + \varepsilon \| y(0) \|^2,$$

and hence the closed-loop system is given by

$$
\left.
\begin{aligned}
\dot{x}(t) &= [A - k(t) B C] x(t) \\
\dot{k}(t) &= 2\varepsilon \langle C x(t), C[A - k(t) B C] x(t) \rangle + \| C x(t) \|^2,
\end{aligned}
\right\} \tag{7.5}
$$

with initial conditions $x(0) = x_0$, $k(0) = \eta_0 + \varepsilon \| y(0) \|^2$. Since the right hand side of (7.5) is continuous and locally Lipschitz in k, x, it follows from classical theory of differential equations that for every $(x_0, k(0)) \in \mathbb{R}^{n+1}$, the closed-loop system has a unique solution $(x(\cdot), k(\cdot)) : [0, \omega) \to \mathbb{R}^{n+1}$, maximally extended over $[0, \omega)$ for some $\omega > 0$. Note that in case of $D = 0$, it is not necessary to assume $\eta(0) \geq 0$.

(b):   We use the notation

$$\eta(t) := \int_0^t \|y(s)\|^2 ds$$

and choose $P = P^T \in \mathbb{R}^{m \times m}$ to be the positive-definite solution of

$$PCB + (CB)^T P = 2I_m.$$

Differentiation of the positive-definite function

$$V(y(t), \eta(t)) := \frac{1}{2}\|y(t)\|_P^2 + \frac{1}{4}\eta(t)^2$$

along (7.5) yields, by using equation (2.8) and $k(t) \geq \eta(t)$, for all $s \in [0, \omega)$,

$$\frac{d}{ds}V(y(s), \eta(s)) \leq M_2\|y(s)\|^2 + M_2\|y(s)\| \left[\|w(s)\| + \|\mathcal{L}(y)(s)\|\right]$$
$$-k(s)\|y(s)\|^2 + \frac{1}{2}\eta(s)\dot{\eta}(s)$$
$$\leq \left[M_2 - \left(1 - \frac{1}{2}\right)\eta(s)\right]\dot{\eta}(s) + M_2\|y(s)\| \left[\|w(s)\| + \|\mathcal{L}(y)(s)\|\right]$$

for some $M_2 > 0$. Integrating this inequality over $[0, t]$, and using (2.9), (2.10) yields

$$V(y(t), \eta(t)) \leq V(y(0), \eta(0)) + \int_0^t \left[M_2 - \frac{1}{2}\eta(s)\right]\dot{\eta}(s)ds$$
$$+ M_2\left[\int_0^t \|y(s)\|^2 ds + M_3^2\|z(0)\|^2\right]$$
$$+ M_2\sqrt{\frac{M_1\|A_2\|\|A_3\|}{\omega}}\int_0^t \|y(s)\|^2 ds$$
$$\leq M_4 + M_4\eta(t) + \int_0^t \left[M_2 - \frac{1}{2}\eta(s)\right]\dot{\eta}(s)ds,$$

for M_3 as in (2.10) and

$$M_4 := V(y(0), \eta(0)) + M_2 M_3^2\|z(0)\|^2 + M_2 + M_2\sqrt{\frac{M_1\|A_2\|\|A_3\|}{\omega}},$$

and hence, by changing variables, for all $t \in [0, \omega)$,

$$V(y(t), \eta(t)) \leq M_4 + M_4\eta(t) + M_2\left[\eta(t) - \eta(0)\right] - \frac{1}{4}\left[\eta(t)^2 - \eta(0)^2\right].$$

If $\eta(\cdot) \notin L_\infty(0, \omega)$, then the right hand side of the above inequality takes negative values, contradicting the positiveness of the left hand side. Thus, $\eta(\cdot) \in L_\infty(0, \omega)$ and therefore, again by the inequality and the definition of $V(y, \eta)$, $y(\cdot) \in L_\infty(0, \omega)$. $\eta(\cdot) \in L_\infty(0, \omega)$ implies $y(\cdot) \in L_2(0, \omega)$, and hence $k(\cdot) \in L_\infty(0, \omega)$.

(c): Boundedness of $k(\cdot)$ yields, by the classical theory of differential equations, $\omega = \infty$. This proves the statements (i), (ii), and (iii) is a consequence of Lemma 2.1.8.

(d): It remains to consider the cases when (II) or (III) are satisfied. We assume (III) is valid and first show that the feedback system is well defined and that there exists a unique solution. Since $W^T[D + D^T]W \geq 0$ and $k(t) \geq 0$ for all $t \geq 0$, it follows as in the proof of Proposition 3.1.4 that

$$I_r + k(t)W^T DW \in GL_r(\mathbb{R}) \quad \text{for all} \quad t \geq 0.$$

Hence, by using the notation

$$\bar{y}(t) = [Z, W]^T y(t), \qquad \bar{y}_1(t) = Z^T y(t), \qquad \bar{y}_2(t) = W^T y(t),$$

and, applying the feedback $u(t) = -k(t)y(t)$ to the second equation in (7.4), we obtain

$$\left.\begin{array}{rcl} \bar{y}_1(t) & = & Z^T Cx(t) \\ \bar{y}_2(t) & = & [I_r + k(t)W^T DW]^{-1} W^T Cx(t). \end{array}\right\} \tag{7.6}$$

The first equation in (7.4) reads now

$$\dot{x}(t) = \left\{ A - B[Z, W] \left[\begin{array}{c} k(t)Z^T \\ (I_r + k(t)W^T DW)^{-1}W^T \end{array} \right] C \right\} x(t).$$

Differentiation of $k(t)$ yields

$$\begin{array}{rcl} \dot{k}(t) & = & 2\varepsilon\langle \bar{y}_1(t), \dot{\bar{y}}_1(t)\rangle + 2\varepsilon\langle \bar{y}_2(t), \dot{\bar{y}}_2(t)\rangle + \|\bar{y}(t)\|^2 \\ & = & 2\varepsilon\langle \bar{y}_1(t), \dot{\bar{y}}_1(t)\rangle + \|\bar{y}(t)\|^2 + 2\varepsilon\langle \bar{y}_2(t), [I_r + k(t)W^T DW]^{-1}W^T C\dot{x}(t)\rangle \\ & & -2\varepsilon\dot{k}(t)\langle \bar{y}_2(t), [I_r + k(t)W^T DW]^{-1}W^T DW\bar{y}_2(t)\rangle. \end{array}$$

In order to get an explicit formula for $\dot{k}(t)$, we shall show that

$$\langle \bar{y}_2(t), [I_r + k(t)W^T DW]^{-1}W^T DW\bar{y}_2(t)\rangle \geq 0. \tag{7.7}$$

To this end let

$$X := W^T DW.$$

Since

$$[I_r + kX]^{-1} X + X^T \left[I_r + kX^T \right]^{-1}$$
$$= [I_r + kX]^{-1} X \left\{ [I_r + kX^T]X^{-T} + X^{-1}[I_r + kX] \right\} X^T \left[I_r + kX^T \right]^{-1}$$
$$= [I_r + kX]^{-1} X \left\{ X^{-T} + X^{-1} + 2kI_r \right\} X^T \left[I_r + kX^T \right]^{-1},$$

it remains to show and $X^{-T} + X^{-1} + 2kI_r \geq 0$ for all $k \geq 0$. This follows from the assumption c) and Remark 3.1.3 (iv), and hence (7.7) is proved.

Therefore, the closed-loop system can be written

$$
\left.
\begin{aligned}
\dot{x}(t) &= \left\{ A - B[Z, W] \begin{bmatrix} k(t)Z^T \\ (I_r + k(t)W^T DW)^{-1}W^T \end{bmatrix} C \right\} x(t) \\
\dot{k}(t) &= \left[1 + 2\varepsilon \langle \bar{y}_2(t), [I_r + k(t)W^T DW]^{-1}W^T DW\bar{y}_2(t) \rangle \right]^{-1} \\
&\quad \cdot \left\{ 2\varepsilon \langle \bar{y}_1(t), \dot{\bar{y}}_1(t) \rangle + \| \bar{y}(t) \|^2 \right. \\
&\quad \left. + 2\varepsilon \langle \bar{y}_2(t), [I_r + k(t)W^T DW]^{-1}W^T C\dot{x}(t) \rangle \right\},
\end{aligned}
\right\} \tag{7.8}
$$

with initial conditions $x(0) = x_0$ and $k(0) = \eta_0 + \varepsilon \|y(0)\|^2$. Inserting (7.6) into the right hand side of (7.8) yields a right hand side which is continuous and locally Lipschitz in k, x, and thus it follows from classical theory of differential equations that the closed-loop system has a unique solution $(x(\cdot), k(\cdot)) : [0, \omega) \to \mathbb{R}^{n+1}$, maximally extended over $[0, \omega)$, for some $\omega > 0$.

(e): By Theorem 3.2.3, the system (7.4) is equivalent to the almost strictly positive real system (3.22) with

$$
\bar{u}(t) = S[Z, W]^T u(t) = -k(t)S\bar{y}(t).
$$

It follows from Lemma 3.1.6, applied to (3.22), that

$$
\frac{d}{dt}V(x(t)) \leq -2\mu V(x(t)) - 2\left[k(t) - \|K\|\right]\|\bar{y}(t)\|^2
$$

for some $K \in \mathbb{R}^{m \times m}$, $\mu > 0$ and positive-definite $V(x) = \langle x, Px \rangle$ as defined in Lemma 3.1.6. Consider the positive-definite function

$$
W(x(t), \eta(t)) \quad := \quad V(x(t)) + \frac{1}{2}\eta(t)^2.
$$

Differentiation of $W(x(s), \eta(s))$, for $s \in [0, \omega)$, along the solution of the feedback system (7.4), $u(t) = -k(t)y(t)$, and using $k(t) \geq \eta(t)$, yields

$$
\begin{aligned}
\frac{d}{ds}W(x(s), \eta(s)) &\leq -2\mu V(x(s)) - 2\left[k(s) - \|K\|\right]\|\bar{y}(s)\|^2 + \eta(s)\dot{\eta}(s) \\
&\leq -2\mu V(x(s)) - 2\left[k(s) - \|K\| - \frac{1}{2}\eta(s)\right]\dot{\eta}(s) \\
&\leq -2\mu V(x(s)) - 2\left[\frac{1}{2}\eta(s) - \|K\|\right]\dot{\eta}(s).
\end{aligned}
$$

Integration of this inequality over $[0, t]$ and changing variables gives

$$
W(x(t), \eta(t)) \leq W(x(0), \eta(0)) - 2\mu \int_0^t V(x(s))ds - 2\int_{\eta(0)}^{\eta(t)} \left[\frac{1}{2}\tau - \|K\|\right]d\tau.
$$

From this inequality we conclude that $x(\cdot) \in L_2(0, \omega)$ and $\eta(\cdot) \in L_\infty(0, \omega)$, since otherwise the right hand side would take negative values, thus contradicting the positiveness of the left hand side. Again, by the inequality and the non-negativeness of $W(x, \eta)$, $W(x(\cdot), \eta(\cdot)) \in L_\infty(0, \omega)$, and hence by the definition of $W(x, \eta)$, $x(\cdot) \in L_\infty(0, \omega)$. The definition of $k(\cdot)$ yields $k(\cdot) \in L_\infty(0, \omega)$, whence, by the classical theory of differential equations, $\omega = \infty$. It now follows from the first equation in (7.8) that $\dot{x}(\cdot) \in L_2(0, \infty)$, and hence Lemma 4.2.1 yields $\lim_{t \to \infty} x(t) = 0$. This proves the statements (i)-(iii) in case (7.4) satisfies (III).

We omit the proof for the case (II) since it is in the same spirit as (d) and (e). This completes the proof. □

Remark 7.1.7
The modification of the gain presented in Theorem 7.1.6 is also applicable to the universal adaptive stabilizer presented in Theorem 5.1.3. That means, for $\varepsilon > 0$ and $\eta_0 \geq 0$, the internal model (5.2) in series with the feedback strategy

$$
\begin{aligned}
e(t) &= y(t) - y_{\text{ref}} \\
v(t) &= -k(t)e(t) \\
k(t) &= \varepsilon \|e(t)\|^2 + \int_0^t \|e(s)\|^2 ds + \eta_0,
\end{aligned}
$$

applied to any system (7.4) satisfying (I)-(III) in Theorem 7.1.6 and $\lim_{s \to \infty} \frac{\alpha(s)}{\beta(s)} > 0$ yields a closed-loop system satisfying assertions (i)-(iv) in Theorem 5.1.3.

Before we shall present a similar gain adaptation modification for the λ-tracking controller of Section 5.2 respectively 6.3, we will discuss and illustrate the effect when the feedback strategy

$$
\left.
\begin{aligned}
u(t) &= -k(t)y(t) \\
k(t) &= \varepsilon \|y(t)\|^2 + \int_0^t \|y(s)\|^2 ds,
\end{aligned}
\right\} \tag{7.9}
$$

is applied to (7.1) for different $\varepsilon > 0$.

Example 7.1.8
In Figure 7.7, $|y(t)|$ and $-k(t)$ is plotted for the closed-loop system (7.1), (7.9) and different ε. For $\varepsilon = 0.01$ (solid line), the behaviour is only slightly better than for $\varepsilon = 0$ as shown in Figure 7.1. But already for $\varepsilon = 0.1$ (dashed line) there is a considerably smaller peak of $y(t)$ and the terminal gain is smaller in magnitude as well. The different behaviour becomes very clear for $\varepsilon = 0.5$ (dotted line): The contribution of $\varepsilon \|y(t)\|^2$ in the gain adaptation ensures faster increase of $k(t)$. As soon as $k(t)$ has reached the stabilizing value 10, in our

example at $t \approx 0.5$, $y(t)$ starts to decrease, and, for $\varepsilon = 0.5$, this yields that $k_\infty \approx 11$, thus considerably closer to the stabilizing gain than for $\varepsilon = 0.01$ or $\varepsilon = 0$. In the case of $\varepsilon = 0.5$, we have $\int_0^{0.5} y(s)^2 ds < 10$ but $\varepsilon y(0.5)$ ensures that $k(0.5) > 10$. For $t > 0.5$, the contribution of $\varepsilon \|y(t)\|^2$ decreases, while the contribution of $\int_0^t y(s)^2 ds$ is increasing. Since $k(t)$ is smaller, the system has a smaller exponential decay rate and thus $y(t)$ needs a longer time to converge to 0.

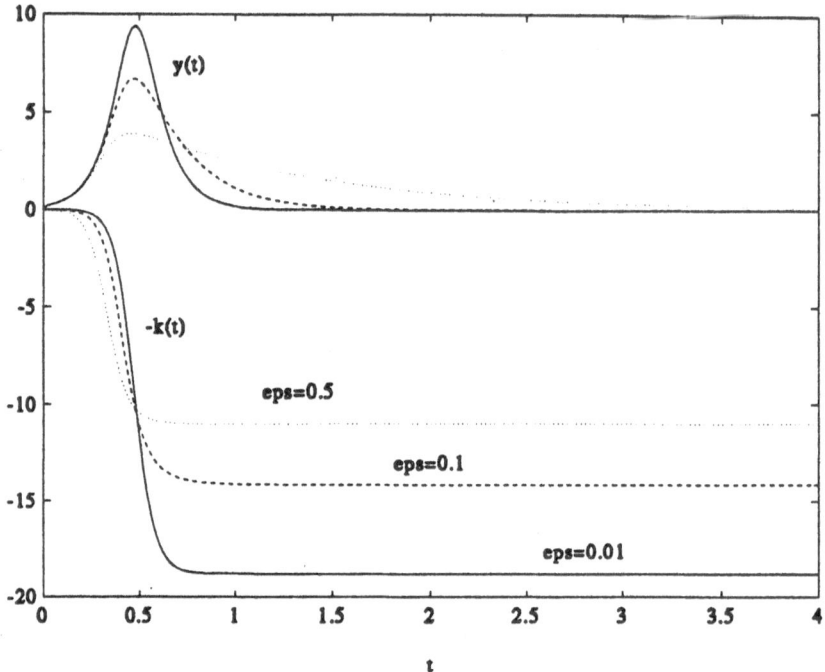

Figure 7.7: Stabilization via $u(t) = -k(t)y(t)$, $k(t) = \varepsilon y(t)^2 + \int_0^t y(s)^2 ds$

Example 7.1.9
In Figure 7.8, we have plotted the output $y(t)$ and $-k(t)$ for the closed-loop system (7.1), (7.9) and $\varepsilon = 0.01, 0.1, 0.25, 0.5, 1, 2, 30$. Note that for $\varepsilon > 0.25$ there is little change in the limiting gain k_∞. However, for large ε, the peak of $y(t)$ is very little, the time in which $k(t)$ reaches $k(t) = 10$ is very small, but the final time until $y(t)$ is close to 0 is very long.

It is also possible to improve the gain and output behaviour of the λ-tracking controller presented in Section 5.2, by modifying the gain adaptation in a similar manner as in Theorem 7.1.6. In the following theorem we shall consider this modified gain adaptation if applied to a nonlinearly perturbed system of the form (6.20).

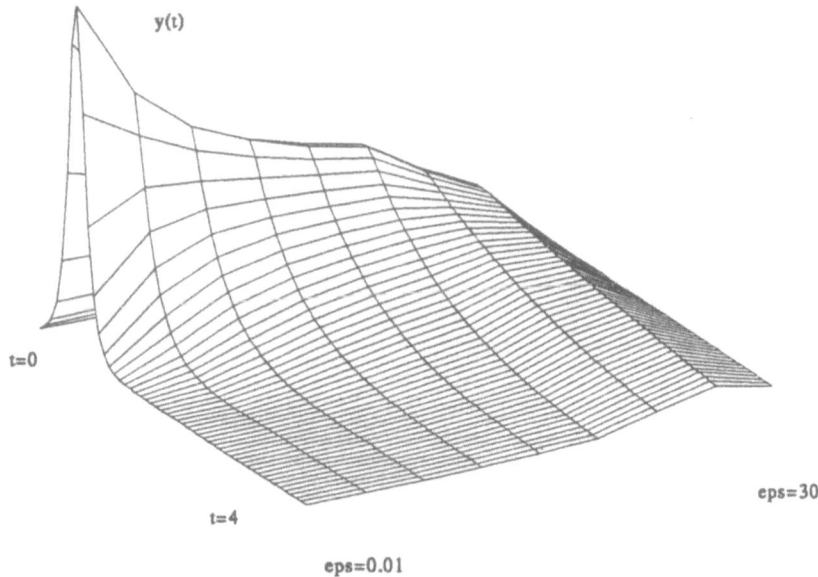

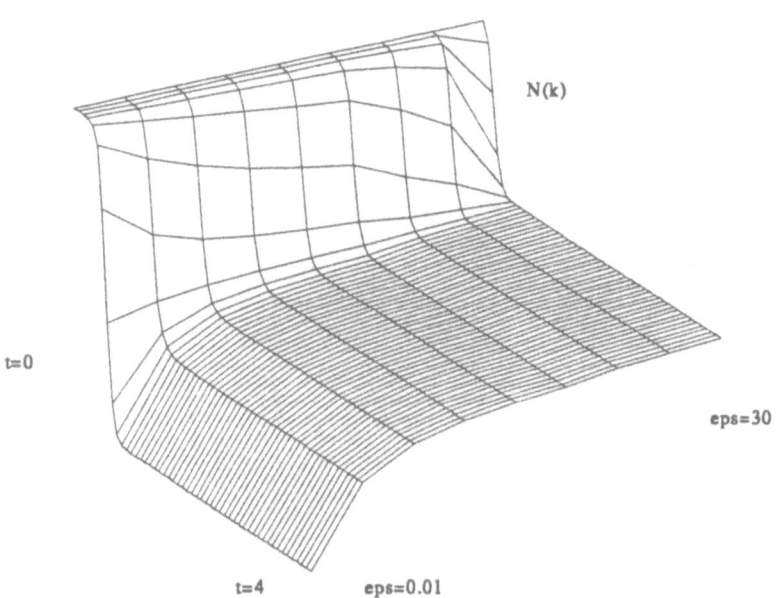

Figure 7.8:Stabilization via $u(t) = -k(t)y(t)$, $k(t) = \varepsilon y(t)^2 + \int_0^t y(s)^2 ds$ and
$\varepsilon = 0.01, 0.1, 0.25, 0.5, 1, 2, 30$

Theorem 7.1.10

Consider the nonlinearly perturbed system

$$
\left.\begin{array}{rcl}
\dot{x}(t) &=& Ax(t) + g(t, x(t)) + B\left[u(t) + f(t, x(t))\right] \quad , x(0) = x_0 \\
y(t) &=& Cx(t) + n(t)
\end{array}\right\} \quad (7.10)
$$

with $(A, B, C) \in \mathbb{R}^{n \times n} \times \mathbb{R}^{n \times m} \times \mathbb{R}^{m \times n}$ minimum phase and $\sigma(CB) \subset \mathbb{C}_+$. Let $g(\cdot, \cdot)$ and $f(\cdot, \cdot)$ be Carathéodory functions satisfying (6.21), $\tilde{f} : \mathbb{R}^m \rightarrow [0, \infty)$ a known differentiable function satisfying (6.22) with derivative having the Carathéodory properties, and $n(\cdot) \in \mathcal{W}^{1, \infty}(\mathbb{R}, \mathbb{R}^m)$.

If $\varepsilon, \lambda > 0$, and $s_\lambda(\cdot) : \mathbb{R}^m \rightarrow [0, \lambda^{-1}]$ satisfying (6.23), then the adaptive feedback mechanism

$$
\left.\begin{array}{rcl}
e(t) &=& y(t) - y_{\text{ref}}(t), \\
u(t) &=& -k(t)\left[1 + \tilde{f}\left(e(t) + y_{\text{ref}}(t)\right) s_\lambda(e(t))\right] e(t) \\
k(t) &=& \varepsilon \tfrac{1}{2} d_\lambda(e(t))^2 \left[\|e(t)\| + \tilde{f}\left(e(t) + y_{\text{ref}}(t)\right)\right] \\
&& + \int_0^t d_\lambda(e(s)) \left[\|e(s)\| + \tilde{f}\left(e(s) + y_{\text{ref}}(s)\right)\right] ds + \eta_0
\end{array}\right\} \quad (7.11)
$$

applied to (7.10), for arbitrary $\eta_0 \geq 0$, $x_0 \in \mathbb{R}^n$, $k_0 \in \mathbb{R}$, $y_{\text{ref}}(\cdot) \in \mathcal{W}^{1, \infty}(\mathbb{R}, \mathbb{R}^m)$, yields a solution $(x(\cdot), k(\cdot)) : [0, \omega) \rightarrow \mathbb{R}^{n+1}$ of the closed-loop system, for some $\omega > 0$, and every solution has on its maximal interval of existence $[0, \omega)$ the properties

(i) $\omega = \infty$,

(ii) $\lim_{t \to \infty} k(t) = k_\infty$ exists and is finite,

(iii) $x(\cdot), k(\cdot) \in L_\infty(0, \infty)$,

(iv) the error $e(t)$ approaches the closed ball $\bar{B}_\lambda(0)$ as $t \to \infty$.

Proof: (a): In order to show existence of a solution we rewrite the closed-loop system as a differential equation. Differentiation of $k(t)$ yields

$$
\begin{aligned}
\dot{k}(t) &= \varepsilon d_\lambda(e(t)) \frac{\langle e(t), \dot{e}(t) \rangle}{\|e(t)\|} \left[\|e(t)\| + \tilde{f}\left(e(t) + y_{\text{ref}}(t)\right)\right] \\
&\quad + \varepsilon \frac{1}{2} d_\lambda(e(t))^2 \left[\frac{\langle e(t), \dot{e}(t) \rangle}{\|e(t)\|} + \tilde{f}\left(e(t) + y_{\text{ref}}(t)\right)\left(\dot{e}(t) + \dot{y}_{\text{ref}}(t)\right)\right] \\
&\quad + d_\lambda(e(t)) \left[\|e(t)\| + \tilde{f}\left(e(t) + y_{\text{ref}}(t)\right)\right].
\end{aligned}
$$

It now follows from (6.24) that the closed-loop system can be written as a inition value problem

$$\frac{d}{dt}\left(e(t), z(t)^T, k(t)\right)^T = F\left(t, e(t), z(t), k(t)\right)$$

with

$$
\begin{aligned}
e(0) &= Cx_0 + n(0) - y_{\text{ref}}(0), \\
z(0) &= Nx_0, \\
k(0) &= \varepsilon d_\lambda(e(0))
\end{aligned}
$$

and $F(\cdot,\cdot) : [0,\infty) \times \mathbb{R}^{n+1} \to \mathbb{R}^{n+1}$ a Carathéodory function. Therefore, existence of a $(x(\cdot), k(\cdot)) : [0,\omega) \to \mathbb{R}^{n+1}$, for some $\omega > 0$, follows.

(b):   We shall establish boundedness of $k(\cdot)$. Set

$$\eta(t) := \int_0^t d_\lambda(e(s)) \left[\|e(s)\| + \tilde{f}\left(e(s) + y_{\text{ref}}(s)\right)\right] ds$$

and

$$W_{p\lambda}(e,\eta) := V_{p\lambda}(e) + \frac{1}{2\alpha}\eta^2,$$

where $\alpha > 0$ will be specified later and $V_{p\lambda}(e)$ is defined as in Lemma 6.3.1. Suppose $\eta(\cdot) \notin L_\infty(0,\omega)$. Using the notation of the proof of Theorem 6.3.3, it is easy to see that, for all $t \in [t_0 \in \omega)$,

$$
\begin{aligned}
W_{p\lambda}(e(t),\eta(t)) &\leq W_{p\lambda}(e(t_0),\eta(t_0)) \\
&\quad + \int_{t_0}^t \left[M(M_3+1) - q^{-1}\gamma k(s)\right] \\
&\qquad\qquad D_{p\lambda}(e(s)) \left[\|e(s)\| + \tilde{f}\left(e(s) + y_{\text{ref}}(s)\right)\right] ds \\
&\quad + \frac{1}{\alpha} \int_{t_0}^t \eta(s)\dot{\eta}(s)ds.
\end{aligned}
$$

Since

$$k(t) \geq \eta(t) \quad \text{for all} \quad t \in [0,\omega)$$

and $\eta(\cdot)$ is assumed to be unbounded, there exists $t_1 \in [t_0,\omega)$ so that

$$M(M_3+1) - q^{-1}\gamma k(s) \leq M(M_3+1) - q^{-1}\gamma\eta(s) < 0 \quad \text{for all} \quad s \in [t_1,\omega)$$

and by Lemma 5.2.3 (i) we conclude

$$W_{p\lambda}(e(t),\eta(t)) \leq M_6 + \int_{t_1}^t p\left[M(M_3+1) - \left(q^{-1} - (\alpha p)^{-1}\right)\eta(s)\right]\dot{\eta}(s)ds.$$

$$(7.12)$$

Choose α sufficiently large so that $q^{-1}\gamma - (\alpha p)^{-1} > 0$ and boundedness of $\eta(\cdot)$ yields that the right hand side of (7.12) takes negative values, thus contradicting the non-negativeness of the left hand side. This proves $\eta(\cdot) \in L_\infty(0,\omega)$.

(c):   Boundedness of $\eta(\cdot)$ yields, by (7.12), $W_{p\lambda}(e(\cdot), \eta(\cdot)) \in L_\infty(0,\omega)$, and hence $V_{p\lambda}(e(\cdot)) \in L_\infty(0,\omega)$, whence $e(\cdot) \in L_\infty(0,\omega)$, and thus, finally, $k(\cdot) \in L_\infty(0,\omega)$.

Similar as in part (d) of the proof of Theorem 6.3.2 we may conclude the assertions (i)-(iii).

(d):   By part (d) of the proof of Theorem 6.3.3 we obtain, for almost all $t \geq 0$,

$$\frac{d}{dt}W_{q\lambda}(e(t),\eta(t)) \leq M_5 q d_\lambda(e(t))\|e(t)\| + \alpha^{-1}\eta(t)\dot\eta(t) \leq M_6\dot\eta(t),$$

where

$$M_6 := \text{ess sup}_{t\geq 0}\left\{M_5 q + \alpha^{-1}\eta(t)\right\}.$$

Hence for

$$U(e,\eta) := W_{q\lambda}(e,\eta) - (M_6 + 1)\eta$$

we have, for almost all $t \geq 0$,

$$\frac{d}{dt}U(e(t),\eta(t)) \leq -\dot\eta(t) = -d_\lambda(e(t))\left[\|e(t)\| + \tilde f(e(t) + y_{\text{ref}}(t))\right].$$

Since $(e(\cdot), x(\cdot), k(\cdot), \eta(\cdot))$ is bounded we may apply LaSalle's Invariance Principle for non-autonomous systems, see LaSalle (1976). This proves the ω-limit set of the bounded solution $(e(\cdot), x(\cdot), k(\cdot), \eta(\cdot))$ is contained in $\{(e, x, k, \eta) \in \mathbb{R}^{n+3} \,|\, |e| \leq \lambda\}$. This proves (iv) and completes the proof. (For a direct proof see part (e) of the proof of Theorem 6.3.2.)   □

If single-input, single-output systems with sector bounded input and output nonlinearities as in Theorem 6.3.4 are considered, then it is possible to prove a similar modification as in Theorem 7.1.10. This is omitted.

7.2   Arbitrary good transient and steady-state response

The main result of this section is as follows: for arbitrary small prespecified $\varepsilon, \delta, T > 0$, we design a feedback controller

$$\begin{aligned} u(t) &= -k(t) \cdot f(y(t), \dot y(t)) \\ \dot k(t) &= g(y(t), \dot y(t)) \end{aligned}$$

with piecewise continuous $f, g : \mathbb{R}^2 \to \mathbb{R}$ (both depending on ε, δ, T), so that, if applied to an arbitrary linear single-input, single-output system with positive high-frequency gain, the overshoot of the output is less than δ, i.e.

$$|y(t)| \leq |y(0)| + \delta \quad \text{for all} \quad t \geq 0,$$

and after time T, the output is within an interval of radius ε, i.e.

$$|y(t)| \le \varepsilon \quad \text{for all} \quad t \ge T,$$

and $\lim_{t \to \infty} x(t) = 0$.
The price which is paid for this result is that the feedback mechanism invokes the derivative of the output.

Before we shall show this result, we present a simpler controller, which guarantees an arbitrarily small output overshoot. This also gives an intuition how to design a controller for the control objective described above.

Derivative Feedback Controller

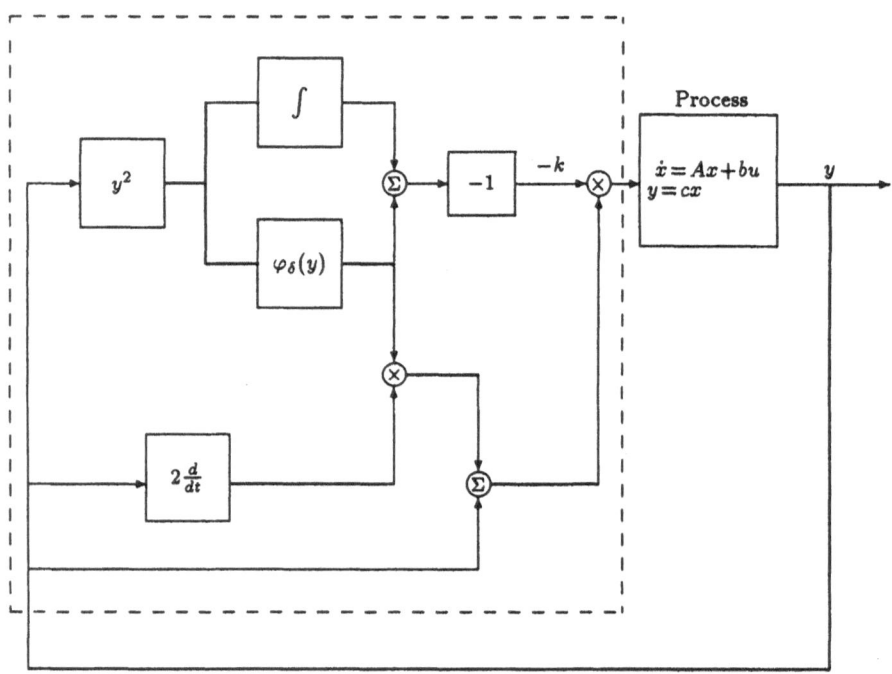

Figure 7.9: Universal adaptive stabilizer with prespecified overshoot

To achieve that $|y(t)|$ does not become larger than the threshold $|y(0)| + \delta$, the idea is to increase the gain if $|y(t)|$ is getting closer to the threshold, thereby forcing $|y(t)|$ to decrease. To this end let

$$k(t) = \eta_0 + \int_0^t y(s)^2 ds + \varphi_\delta(y)(t)$$

with

$$\varphi_\delta(y)(t) := \frac{1}{(|y(0)| + \delta)^2 - y(t)^2} \tag{7.13}$$

for some $\delta > 0$ and $\eta_0 \geq 0$. Note that $\varphi_\delta(y)(t)$ tends to $+\infty$ if $|y(t)| \to |y(0)| + \delta$.

Theorem 7.2.1
Suppose the system

$$\begin{aligned}
\dot{x}(t) &= Ax(t) + bu(t) &, x(0) \in \mathbb{R}^n \\
y(t) &= cx(t),
\end{aligned} \left.\right\} \tag{7.14}$$

with $(A, b, c) \in \mathbb{R}^{n \times n} \times \mathbb{R}^n \times \mathbb{R}^{1 \times n}$ is minimum phase and $cb > 0$. Let $\delta > 0$ and $\varphi_\delta(y)(t)$ be given as in (7.13). If the feedback strategy

$$\begin{aligned}
u(t) &= -k(t)\left[y(t) + 2\varphi_\delta(y)(t)^2 \dot{y}(t)\right] \\
k(t) &= \eta_0 + \int_0^t y(s)^2 ds + \varphi_\delta(y)(t)
\end{aligned} \left.\right\} \tag{7.15}$$

is applied to (7.14), for arbitrary $\eta_0 \geq 0$, $x_0 \in \mathbb{R}^n$, then the closed-loop system has the properties

(i) the unique solution $(x(\cdot), k(\cdot)) : [0, \infty) \to \mathbb{R}^{n+1}$ exists,

(ii) $\lim_{t \to \infty} k(t) = k_\infty$ exists and is finite,

(iii) $x(\cdot) \in L_2(0, \infty) \cap L_\infty(0, \infty)$ and $\lim_{t \to \infty} x(t) = 0$.

(iv) $|y(t)| \leq |y(0)| + \delta$ for all $t \geq 0$.

Proof: (a): First, local existence and uniqueness of the solution $(x(\cdot), k(\cdot))$ of the closed-loop system is shown. Without loss of generality, we may assume that (7.14) is of the form (2.3). Therefore, the closed-loop system is given by

$$\dot{y}(t) = A_1 y(t) + A_2 z(t) - cbk(t)\left[y(t) + 2\varphi_\delta(y)(t)^2 \dot{y}(t)\right] , y(0) \in \mathbb{R}$$
$$\dot{z}(t) = A_3 y(t) + A_4 z(t) , z(0) \in \mathbb{R}^{n-1}$$
$$\dot{k}(t) = y^2(t) + 2\varphi_\delta(y)(t)^2 \dot{y}(t)y(t) , k(0) = \eta_0 + \varphi_\delta(y)(0)^2.$$

Since $k(t) \geq 0$, as long as it exists, the initial value problem is equivalent to

$$\left.\begin{aligned}
\dot{y}(t) &= \left[1 + 2cb\varphi_\delta(y)(t)^2 k(t)\right]^{-1} \{(A_1 - cbk(t))y(t) + A_2 z(t)\} \\
\dot{z}(t) &= A_3 y(t) + A_4 z(t) \\
\dot{k}(t) &= y(t)^2 + \frac{2\varphi_\delta(y)(t)^2 y(t)}{1 + 2cb\varphi_\delta(y)(t)^2 k(t)} \{(A_1 - cbk(t))y(t) + A_2 z(t)\}
\end{aligned}\right\} \tag{7.16}$$

with initial values $y(0) \in \mathbb{R}, z(0) \in \mathbb{R}^{n-1}, k(0) = \eta_0 + \varphi_\delta(y)(0)^2$. It is easy to see that the right hand side of the differential equation above is continuous and Lipschitz at each $(y, z, k) \in \mathbb{R} \times \mathbb{R}^{n-1} \times [0, \infty)$. Therefore, it follows from the theory of ordinary differential equations that there exists a unique solution $(x(\cdot), k(\cdot)) : [0, \omega) \to \mathbb{R}^{n+1}$, maximally extended over $\omega \in (0, \infty]$.

(b): It will be shown that $k(\cdot) \in L_\infty(0, \omega)$. Inserting $u(t)$ into the inequality given in Lemma 2.1.6 for $P = 1$ yields, for some $M > 0$ and all $t \in [0, \omega)$,

$$
\frac{1}{2}y(t)^2 \;\leq\; M\|x(0)\|^2 + M\int_0^t y(s)^2 ds - cb\int_0^t k(s)y(s)\left[y(s) + 2\varphi_\delta(y)(s)^2\dot{y}(s)\right]ds
$$

$$
\leq\; M\|x(0)\|^2 + M[k(t) - k(0)] - cb\int_0^t k(s)\dot{k}(s)ds
$$

$$
\leq\; M\|x(0)\|^2 + M[k(t) - k(0)] - cb\int_{k(0)}^{k(t)} \mu\, d\mu
$$

$$
=\; M\|x(0)\|^2 + [k(t) - k(0)]\left[M - \frac{cb}{2}(k(t) + k(0))\right].
$$

If $k(\cdot) \notin L_\infty(0, \omega)$, then the right hand side tends to $-\infty$, thus contradicting the non-negativeness of the left hand side. This proves $k(\cdot) \in L_\infty(0, \omega)$.

(c): By definition of $k(\cdot)$ and the inequality in (b), boundedness of $k(\cdot)$ yields $y(\cdot) \in L_2(0, \omega) \cap L_\infty(0, \omega)$. Since A_4 is exponentially stable, (3.5) yields $z(\cdot) \in L_2(0, \omega) \cap L_\infty(0, \omega)$. It follows from (7.16) that $\omega = \infty$. This proves (i), (ii), and the first statement in (iii). The proof of the second statement in (iii) is very similar to the proof of Lemma 2.1.8. (iv) follows from boundedness of $k(\cdot)$.
This completes the proof.   □

The control strategy (7.15) will now be extended in order to achieve all three control objectives described in the introduction.

Theorem 7.2.2
Suppose the system

$$
\left.
\begin{aligned}
\dot{x}(t) &= Ax(t) + bu(t) \quad, x(0) \in \mathbb{R}^n \\
y(t) &= cx(t),
\end{aligned}
\right\}
\tag{7.17}
$$

with $(A, b, c) \in \mathbb{R}^{n \times n} \times \mathbb{R}^n \times \mathbb{R}^{1 \times n}$ is minimum phase and $cb > 0$. Let $\delta, \varepsilon, T > 0$

be given. If the adaptive feedback strategy

$$
u(t) \;=\; \begin{cases} -k(t)\left[1+\dfrac{1}{(T-t)^2}+2\dfrac{1}{T-t}+\dfrac{\dot y(t)}{y(t)}\right]y(t) & ,\ \text{if}\ \ t\in[0,t^*) \\[2mm] -k(t)\left[y(t)+2\varphi_\varepsilon^*(y)(t)^2\dot y(t)\right] & ,\ \text{if}\ \ t\ge t^* \end{cases}
$$

$$
k(t) \;=\; \eta_0+\int_0^t y(s)^2\,ds + \begin{cases} \dfrac{1}{T-t}y(t)^2 & ,\ \text{if}\ \ t\in[0,t^*) \\[2mm] \varphi_\varepsilon^*(y)(t) & ,\ \text{if}\ \ t\ge t^* \end{cases}
$$

with

$$
t^* := \min\{t\in[0,T)\,|\,|y(t)|=\tfrac{\varepsilon}{2}\},
$$

and

$$
\varphi_\varepsilon^*(y)(t) := \frac{1}{(|y(t^*)|+\varepsilon)^2 - y(t)^2},
$$

is applied to (7.17), for arbitrary $\eta_0\ge 0$, $x_0\in\mathbb{R}^n$, then the closed-loop system has the properties

(i) the unique solution $(x(\cdot),k(\cdot)):[0,\infty)\to\mathbb{R}^{n+1}$ exists,

(ii) $\lim_{t\to\infty} k(t)=k_\infty$ exists and is finite,

(iii) $x(\cdot)\in L_2(0,\infty)\cap L_\infty(0,\infty)$ and $\lim_{t\to\infty} x(t)=0$,

(iv) $|y(t)|\le|y(0)|+\delta$ for all $t\in[0,T]$,

(v) $|y(t)|\le\varepsilon$ for all $t\ge T$.

Proof: (a): Without loss of generality, we may assume that (7.17) is of the form (2.3). Differentiation of $k(\cdot)$ yields that the initial value problem can be written

$$
\begin{aligned}
\dot y(t) &= [1+2cbk(t)]^{-1}\left\{A_1 y(t)+A_2 z(t)-cbk(t)\left(1+\tfrac{1}{(T-t)^2}\right)y(t)\right\}\\
\dot z(t) &= A_3 y(t)+A_4 z(t)\\
\dot k(t) &= \left[1+\tfrac{1}{(T-t)^2}\right]y(t)^2+\tfrac{2}{T-t}y(t)\dot y(t)
\end{aligned}
$$

with initial conditions $y(0)\in\mathbb{R}$, $z(0)\in\mathbb{R}^{n-1}$, $k(0)=\eta_0+\tfrac{1}{T}y(0)^2$. It is easy to see that the right hand side of the differential equation above is continuous and Lipschitz in $(y,z,k)\in\mathbb{R}\times\mathbb{R}^{n-1}\times[0,\infty)$ and measurable and locally integrable in $t\in[0,T)$. Therefore, it follows from the theory of ordinary differential equations that there exists a unique solution $(x(\cdot),k(\cdot)):[0,\omega)\to\mathbb{R}^{n+1}$, maximally extended over $\omega\in(0,\infty]$.

(b): It will be shown that $k(\cdot)\in L_\infty(0,\omega)$. Inserting the feedback into the inequality given in Lemma 2.1.6 for $P=1$ yields, for some $M>0$ and all

$t \in [0, \omega)$,

$$\frac{1}{2} y(t)^2 \;\leq\; M\|x(0)\|^2 + M \int_0^t y^2(s) ds - cb \int_0^t k(s) \dot{k}(s) ds$$

$$= \; M\|x(0)\|^2 + [k(t) - k(0)] \left[M - \frac{cb}{2}[k(t) + k(0)] \right].$$

If $y(\cdot)$ or $k(\cdot) \notin L_\infty(0, \omega)$, then the right hand side tends to $-\infty$, thus contradicting the non-negativeness of the left hand side. Thus there is not a finite escape time at any $t' \in [0, \omega)$. The existence of t^* also is a consequence of the inequality and the construction of $k(t)$.

(c): In a similar manner as in part (a) of the proof of Theorem 7.2.1, we conclude that the closed-loop system for $t \geq t^*$ is given by (7.16) with initial conditions $y(t^*) \in \mathbb{R}$, $z(t^*) \in \mathbb{R}^{n-1}$, $k(t^*) \geq 0$, and φ_δ replaced by φ_t^*. As before, existence and uniqueness of a solution $(x(\cdot), k(\cdot)) : [t^*, t^* + \omega) \to \mathbb{R}^{n+1}$, maximally extended over $\omega \in (0, \infty]$, follows.

(d): The remainder of the proof is omitted, it is completely analogous to part (b) and (a) in the proof of Theorem 7.2.1.
This completes the proof. $\square$

Remark 7.2.3
It is easy to see that the adaptive control strategy in Theorem 7.2.2 can be applied to the universal asymptotic tracking problem as well. More precisely: If in the feedback mechanism in Theorem 7.2.2 $y(t)$ is replace by $e(t) = y(t) - y_{\text{ref}}$, and $u(t)$ by $v(t)$, then the series interconnection with the internal model (5.8), where $\hat{d} > 0$, yields a universal adaptive tracking controller, i.e. the assertions (i)-(iv) in Theorem 5.1.3 are satisfied and, moreover, the error satisfies

$$\|e(t)\| \leq \|e(0)\| + \delta \qquad \text{for all} \quad t \in [0, T],$$
$$\|e(t)\| \leq \varepsilon \qquad\qquad \text{for all} \quad t \geq T.$$

Example 7.2.4
In Figure 7.10, the output $y(t)$ and the gain $k(t)$ is plotted for the switching strategy (7.15) applied to (7.1) for $\eta_0 = 0$ and $\delta = 3.3$. Note the difference to Figure 7.1. The initial value of the output is $y(0) = 0.2$ and by construction of $k(t)$, $y(t)$ cannot meet the threshold $|y(0)| + \delta = 3.5$, however it is very close to it. The boundary of $y(t)$ yields that the contribution of $\int_0^t y(s)^2 ds$ in $k(t)$ is 'almost linear', therefore the system needs a longer time to settle down. The terminal gain k_∞ is close to that in Example 7.1.1.

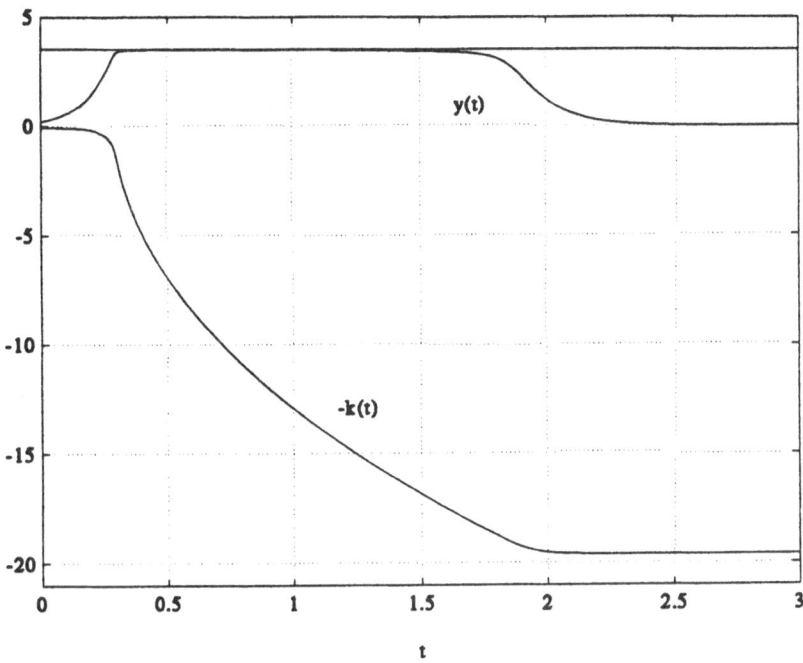

Figure 7.10: Prespecified output overshoot

Example 7.2.5

In Figure 7.11 the effect of different δ in the switching strategy (7.12) applied to (7.1) is illustrated. We vary $\delta = 0.8, 1.3, 1.8, \ldots, 4.8$. In all cases, the system is initially unstable and $y(t)$ reaches the threshold very quickly. The smaller the prespecified overshoot is forced to be, the smaller is the contribution of $\int_0^t y(s)^2 ds$ in $k(t)$ and thus $k(t)$ takes longer to reach its stabilizing value, hence $y(t)$ needs longer to converge to 0. In all cases, k_∞ is approximately the same.

7.3   Notes and References

For the restricted class of scalar systems, the modification of the gain adaptation as in Theorem 7.1.6 has been used in Heymann et al. (1985).

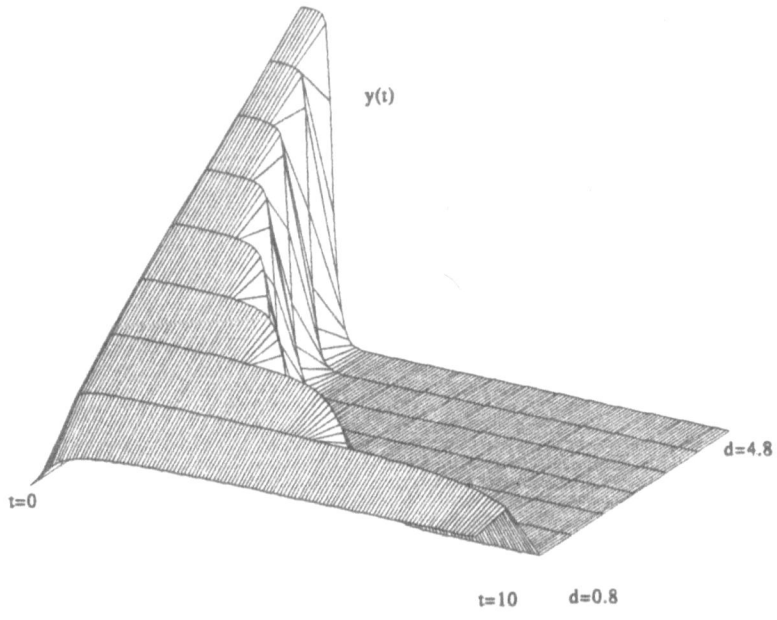

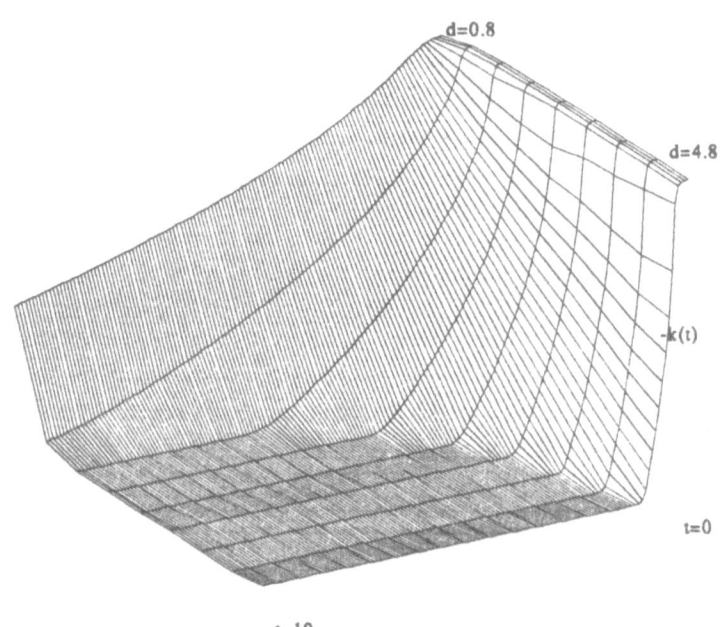

Figure 7.11: Prespecified output overshoot with thresholds

Chapter 8

Exponential Stability of the Terminal System

All results on universal adaptive stabilization or tracking presented in Chapter 4 and Section 5.1 have only ensured that the solution $x(t)$ is decaying asymptotically or exponentially to 0, the terminal system (see Definition 1.1.2) might still be unstable. For scalar systems with known sign of the high-frequency gain the terminal system is exponentially stable, see the closed-loop system (1.5) in the Introduction. However, one can easily construct an example which shows that for single-input, single-output systems (A, b, c) of order $n \geq 2$ with $cb > 0$ the feedback strategy does not lead to an exponentially stable terminal systems. Even in the scalar case, if the sign of the high-frequency gain is unknown,

the terminal system might be $\dot{x}(t) = 0 \cdot x(t)$, see the example in the introduction of Section 4.4. Numerical simulations have shown that in most cases 'almost always' an exponentially stable terminal system is achieved. In this chapter, we shall prove this observation for strictly proper, single-input, single-output minimum phase systems and the piecewise constant feedback strategy introduced in Section 4.5. To this end, firstly the root-loci of closed-loop systems is studied in Section 8.1, and the main result is presented in Section 8.2. That is, for fixed initial conditions k_0, the set of stabilizing thresholds is dense in the space of thresholds under consideration and, at each time when the gain adaptation switches, the probability that the new state trajectory is unstable is 1 as long as the gain is not large enough to stabilize the terminal system.

8.1   Root-loci of minimum phase systems

In this section, we consider the root-loci of the system

$$\begin{aligned}
\dot{x}(t) &= Ax(t) + Bu(t), \qquad x(0) = x_0 \\
y(t) &= Cx(t),
\end{aligned} \Bigg\} \qquad (8.1)$$

under feedback

$$u(t) = -kKy(t)$$

where $(A, B, C) \in \mathbb{R}^{n \times n} \times \mathbb{R}^{n \times m} \times \mathbb{R}^{m \times n}$, $K \in \mathbb{R}^{m \times m}$ fixed, and $k \in \mathbb{R}$ varies. That is, the closed-loop system

$$\dot{x}(t) = A_k x(t), \quad A_k := A - kBKC$$

parametrized by $k \in \mathbb{R}$. Throughout this section, (8.1) is assumed to be minimum phase and most of the results are on single-input, single-output systems.

Remark 8.1.1
Basic properties of the linearly perturbed matrix A_k parametrized by $k \in \mathbb{R}$, no extra assumptions on A, B, C, K, are derived in Kato (1976), Ch. II. There it is shown, that the numbers of eigenvalues of $A_k = A - kBKC$ is a constant l independent of k, with the exception of some *exceptional points*. Theses points originate from the algebraic singularities of the (branches of) the solutions of $det(\lambda I_n - A_k) = 0$. In each compact set of $\mathbb{R}$ there is only a finite number of such exceptional points in k. Let $I \subset \mathbb{R}$ be an interval not containing any exceptional points. Then the eigenvalues $\lambda_i(k)$ of A_k depend analytically on $k \in I$. Moreover, the *total projection* $P_{\lambda_i}(k)$ on the *total eigenspace* associated with $\lambda_i(k)$ is analytic in $k \in I$, and so is the eigennilpotent $D_{\lambda_i}(k)$ which satisfies

$$D_{\lambda_i}(k) = (\lambda_i(k)I_n - A_k) P_{\lambda_i}(k).$$

Since $D_{\lambda_i}(k)$ is analytic in $k \in I$, there exists an analytic vector $w_{\lambda_i}(k) \not\equiv 0$ such that $w_{\lambda_i}(k) \in \ker D_{\lambda_i}(k)$, see e.g. in Gohberg et al. (1982) p.388. Therefore,

$$v_{\lambda_i}(k) := P_{\lambda_i}(k)w_{\lambda_i}(k)$$

is an eigenvector of A_k belonging to $\lambda_i(k)$, and depending analytically on $k \in I$.

The following proposition shows interestingly, that, if k varies, then all unstable eigenvalues of A_k are moving.

Proposition 8.1.2
Suppose the system (8.1) is minimum phase with $\det(CB) \neq 0$, and let $K \in \mathbb{R}^{m \times m}$. If $\lambda_k : I \to \overline{\mathbb{C}}_+$ denotes an analytic parametrization of any unstable

eigenvalue of $A_k = A - kBKC$ on some open interval $I \subset \mathbb{R}$, then $\lambda_k \not\equiv$ constant on I.

Proof: Using the state space transformation of Lemma 2.1.3, we have

$$
|sI_n - A_k| = \begin{vmatrix} sI_m - A_1 + kCBK & -A_2 \\ -A_3 & sI_{n-m} - A_4 \end{vmatrix}
$$
$$
= |sI_{n-m} - A_4| \cdot |sI_m - A_1 + kCBK - A_2(sI_{n-m} - A_4)^{-1}A_3|.
$$

Suppose, for some $\lambda \in \overline{\mathbb{C}}_+$ we have $\lambda_k \equiv \lambda$ for all $k \in I$, i.e.

$$
|\lambda I_n - A_k| = 0 \quad \text{for all} \quad k \in I.
$$

Since, by the minimum phase assumption, $|\lambda I_{n-m} - A_4| \neq 0$, it follows that for

$$
M := \lambda I_m - A_1 - A_2(\lambda I_{n-m} - A_4)^{-1}A_3,
$$

we have

$$
|M + kCBK| = 0 \quad \text{for all} \quad k \in I
$$

or, equivalently,

$$
|\hat{M} + kI_m| = 0 \quad \text{for all} \quad k \in I,
$$

where $\hat{M}$ is a Jordan form of $M(CBK)^{-1}$. Thus,

$$
|\hat{M} + kI_m| = \prod_{i=1}^{m}(\mu_i + k) = 0 \quad \text{for all} \quad k \in I,
$$

where $\mu_1, \ldots, \mu_m$ denote the eigenvalues of $\hat{M}$. Since the μ_i do not depend on k, the last equality yields a contradiction, and the proposition is proved. $\square$

The following proposition shows in particular, that for single-input, single-output, minimum phase systems with positive high-frequency gain $cb > 0$, the matrix A_k has, for all but finitely many k, only *distinct* unstable eigenvalues.

Proposition 8.1.3
If a single-input, single-output, minimum phase system $(A, b, c) \in \mathbb{R}^{n \times n} \times \mathbb{R}^n \times \mathbb{R}^{1 \times n}$ with $cb \neq 0$ is considered, then the set

$$
\{k \in \mathbb{R} \,|\, A_k = A - kbc \quad \text{has eigenvalues in } \overline{\mathbb{C}}_+ \text{ of multiplicity } l \geq 2\}
$$

is discrete in $\mathbb{R}$.
If $cb > 0$, then the above set is finite.

Proof: Let $I \subset \mathbb{R}$ be an open interval excluding exceptional points, see Remark 8.1.1, and let $\lambda_k : I \to \overline{\mathbb{C}}_+$ denote an analytic parametrization of an unstable eigenvalue of A_k. It then remains to show that λ_k has multiplicity 1 almost everywhere. Suppose the contrary, that is for $l \geq 2$ the exists a $q_k(s) \in \mathbb{R}[s]$ such that

$$|sI_n - A_k| = (s - \lambda_k)^l q_k(s), \quad q_k(\lambda_k) \neq 0 \quad \text{for all} \quad k \in I. \qquad (8.2)$$

Inserting (8.2) for $k^* \in I$ into the following well known relationship, see for example Kailath (1980) p.651,

$$|sI_n - A_k| = |sI_n - A_{k^*}| + (k - k^*)z(s), \quad \text{where} \quad z(s) := \begin{vmatrix} sI_n - A_{k^*} & b \\ -c & 0 \end{vmatrix},$$

yields

$$|sI_n - A_k| = (s - \lambda_{k^*})^l q_{k^*}(s) + (k - k^*)z(s). \qquad (8.3)$$

Differentiation of (8.3) gives

$$\frac{d}{ds}|sI_n - A_k| = (s - \lambda_{k^*})^{l-1}[lq_{k^*}(s) + (s - \lambda_{k^*})q'_{k^*}(s)] + (k - k^*)z'(s). \qquad (8.4)$$

Since $det(\lambda_k I_n - A_k) \equiv 0$ on I, and, by the minimum phase assumption, $z(\lambda_k) \neq 0$ for all $k \in I$, we conclude from (8.3) and (8.4) that

$$0 = (\lambda_k - \lambda_{k^*})^{l-1}\psi(\lambda_k), \qquad (8.5)$$

where

$$\psi(\lambda_k) := lq_{k^*}(\lambda_k) + (\lambda_k - \lambda_{k^*})q'_{k^*}(\lambda_k) - \frac{(\lambda_k - \lambda_{k^*})q_{k^*}(\lambda_k)}{z(\lambda_k)}z'(\lambda_k).$$

Since $\psi(\lambda_k^*) \neq 0$, it follows from the continuity of the arguments in the definition of $\psi(\lambda_k)$ that, for sufficiently small $\varepsilon > 0$,

$$\psi(\lambda_k) \neq 0 \quad \text{for all} \quad k \in (k^* - \varepsilon, k^* + \varepsilon).$$

Therefore, (8.5) yields $\lambda_k = \lambda_k^*$ for all $k \in (k^* - \varepsilon, k^* + \varepsilon)$, which contradicts Proposition 8.1.2. This proves the first part of the proposition.

Finiteness of the set, in case of $cb > 0$, follows from the fact that $\sigma(A_k) \subset \mathbb{C}_-$ for all $k \geq k'$, for some $k' > 0$, see Remark 2.2.5. This completes the proof. $\square$

The following lemma shows that the projection of each fixed nonzero vector $\zeta \in \mathbb{R}^n$ onto the k-depending unstable subspace of A_k is nonzero, with the exception of discrete points $k \in \mathbb{R}$.

Lemma 8.1.4
Suppose the system $(A, b, c) \in \mathbb{R}^{n \times n} \times \mathbb{R}^n \times \mathbb{R}^{1 \times n}$ is minimum phase and controllable. Let $I \subset \mathbb{R}$ be an open interval such that $\lambda_k : I \rightarrow \overline{\mathbb{C}}_+$ is

an analytic parametrization of an unstable eigenvalue of $A_k = A - kbc$ with eigenvector v_k. Then, for $\zeta \in \mathbb{R}^n$, we have

$$\langle v_k, \zeta \rangle = 0 \quad \text{for all} \quad k \in I \quad \Longleftrightarrow \quad \zeta = 0.$$

Proof: Suppose $\langle v_k, \zeta \rangle = 0$ for all $k \in I$. Since

$$\begin{bmatrix} \lambda_k I_n - A_k & b \\ \zeta^T & 0 \end{bmatrix} \begin{bmatrix} I_n & 0 \\ -kc & I_m \end{bmatrix} = \begin{bmatrix} \lambda_k I_n - A & b \\ \zeta^T & 0 \end{bmatrix},$$

it follows from the controllability assumption that

$$n \le rk \begin{bmatrix} \lambda_k I_n - A & b \\ \zeta^T & 0 \end{bmatrix} = rk \begin{bmatrix} \lambda_k I_n - A_k & b \\ \zeta^T & 0 \end{bmatrix}. \tag{8.6}$$

Since v_k is a right eigenvector, we have

$$\begin{bmatrix} \lambda_k I_n - A_k & b \\ \zeta^T & 0 \end{bmatrix} \begin{pmatrix} v_k \\ 0 \end{pmatrix} = 0$$

and (8.6) yields

$$rk \begin{bmatrix} \lambda_k I_n - A & b \\ \zeta^T & 0 \end{bmatrix} = n \quad \text{for all} \quad k \in I. \tag{8.7}$$

Using the identity theorem of analytic functions and the fact that λ_k is not constant, see Proposition 8.1.2, (8.7) yields

$$rk \begin{bmatrix} s I_n - A & b \\ \zeta^T & 0 \end{bmatrix} = n \quad \text{for all} \quad s \in \mathbb{C}, \tag{8.8}$$

and hence the $(n+1) \times (n+1)$ matrix in (8.8) is singular over the field $\mathbb{R}(s)$. Thus there exists a nonzero pair $(\varphi(\cdot), \alpha(\cdot)) \in \mathbb{R}[s]^n \times \mathbb{R}[s]$ such that

$$\varphi(s)^T (s I_n - A) = \alpha(s) \zeta^T \quad \text{and} \quad \varphi(s)^T b = 0 \quad \text{for all} \quad s \in \mathbb{C}. \tag{8.9}$$

If, for every nonzero pair $(\varphi(\cdot), \alpha(\cdot))$ satisfying (8.9), it holds that $\alpha(\cdot) \equiv 0$, then

$$\varphi(s)^T [s I_n - A, b] \equiv 0,$$

and, by right invertibility of $[s I_n - A, b]$, $\varphi(\cdot) \equiv 0$, which contradicts $(\varphi(\cdot), \alpha(\cdot)) \not\equiv 0$. Therefore, there exists a pair with $\alpha(\cdot) \not\equiv 0$. Considering $(s I_n - A)^{-1}$ as an element of $\mathbb{R}(s)^{n \times n}$, we obtain

$$0 = \alpha(s) \zeta^T (s I_n - A)^{-1} b,$$

and hence

$$0 = \zeta^T \sum_{i=0}^{\infty} s^{-(i+1)} A^i b,$$

which yields

$$0 = \zeta^T \left[b, Ab, \ldots, A^{n-1} b \right].$$

Since (A, b) is controllable, the controllability matrix is right invertible, which implies that $\zeta = 0$. This proves the lemma.   □

The controllability assumption in Lemma 8.1.4 is essential. Lemma 8.1.4 does not hold true for systems $(A, b, c) \in \Sigma$ which are not controllable. Consider for example

$$A = \begin{bmatrix} 0 & 0 \\ 0 & -1 \end{bmatrix}, \quad c = (1, 0), \quad b = \begin{pmatrix} 1 \\ 0 \end{pmatrix}, \quad \text{and} \quad A_k = \begin{bmatrix} -k & 0 \\ 0 & -1 \end{bmatrix}.$$

For $k \in (-\infty, 0)$, the right eigenvector corresponding to $\lambda_k = -k$ is $v_k = (1, 0)^T$. However, for $\zeta = (0, 1)^T$ we obtain $\langle v_k, \zeta \rangle = 0$ for all $k \in (-\infty, 0)$.

Before the main result of this section is proved, a technical lemma is required.

Lemma 8.1.5
If $A \in \mathbb{R}^{n \times n}$, $x_0 \in \mathbb{R}^n$, and $\lambda \in \overline{\mathbb{C}}_+$ so that

$$Av = \lambda v \quad \text{and} \quad \langle v, x_0 \rangle \neq 0,$$

then the solution of the initial value problem

$$\dot{x}(t) = Ax(t), \quad x(0) = x_0$$

is such that $e^{At} x_0 \nrightarrow 0$ as $t \to \infty$.

Proof: The solution $x(t)$ can be written

$$x(t) = e^{At} x_0 = \alpha(x_0) e^{\lambda t} v + \tilde{x}(t),$$

where $\tilde{x}(t)$ is linearly independent of $e^{\lambda t} v$. $\alpha(x_0)$ denotes the coordinate of the projection of x_0 on $v\mathbb{R}$, which is given by

$$\alpha(x_0) = v(v^T v)^{-1} v^T x_0 = \frac{1}{\|v\|^2} v^T x_0.$$

Now $\langle v, x_0 \rangle \neq 0$ yields $\alpha(x_0) \neq 0$, and the result follows since λ is an unstable eigenvalue.   □

We are now in a position to prove the main result of this section, which, in particular, says the following. Suppose (A, b, c) is a controllable minimum phase system such that the closed-loop system $\dot{x}(t) = [A - kbc]x(t)$ is exponentially stable for k sufficiently large. Let $I \subset [0, \infty)$ be a finite union of closed intervals so that $A_k = A - kbc$ is exponentially stable whenever $k \in [0, \infty) \setminus I$. We obtain the result that, if $x_0 \in \mathbb{R}$ is fixed, then the solution of $\dot{x}(t) = A_k x(t)$, $x(0) = x_0$, is unstable for all but finitely many $k \in I$.

Theorem 8.1.6
If the system

$$\begin{aligned}
\dot{x}(t) &= Ax(t) + bu(t), \qquad x(0) = x_0 \\
y(t) &= cx(t),
\end{aligned}$$

is controllable, minimum phase and of relative degree 1, and $x_0 \in \mathbb{R}^n$, $x_0 \neq 0$, is fixed, then the solution of the initial value problem

$$\dot{x}(t) = [A - kbc]x(t), \quad x(0) = x_0$$

satisfies

(i) The set

$$\left\{ k \in \mathbb{R} \, | \, \sigma(A - kbc) \cap \overline{\mathbb{C}}_+ \neq \emptyset \quad \text{and} \quad \lim_{t \to \infty} e^{[A-kbc]t} x_0 = 0 \right\}$$

is discrete in $\mathbb{R}$.

(ii) If $cb > 0$, then the set

$$\left\{ k \geq 0 \, | \, \sigma(A - kbc) \cap \overline{\mathbb{C}}_+ \neq \emptyset \quad \text{and} \quad \lim_{t \to \infty} e^{[A-kbc]t} x_0 = 0 \right\}$$

is finite.

Proof: Let $\mathcal{E}$ denote the discrete set of exceptional points k defined in Remark 8.1.1. (i) is proved if it can be shown that the set

$$\left\{ k \in \mathbb{R} \setminus \mathcal{E} \, | \, \sigma(A_k) \cap \overline{\mathbb{C}}_+ \neq \emptyset \quad \text{and} \quad \lim_{t \to \infty} e^{A_k t} x_0 = 0 \right\}$$

is discrete. Let $\mathbb{R} \setminus \mathcal{E} = \bigcup_{i \in \mathbb{N}} I_i$ be the countable union of disjoint open intervals. It remains to prove that for every $i \in \mathbb{N}$ the set

$$\left\{ k \in I_i \, | \, \sigma(A_k) \cap \overline{\mathbb{C}}_+ \neq \emptyset \quad \text{and} \quad \lim_{t \to \infty} e^{A_k t} x_0 = 0 \right\}$$

is discrete. Let $I := I_{i_0}$ for some $i_0 \in \mathbb{N}$, and $\lambda_k : I \to \overline{\mathbb{C}}_+$, $v_k : I \to \mathbb{C}^n$ denote analytic parametrizations of an unstable eigenvalue - eigenvector pair. It follows from Lemma 8.1.4 that the analytic map $k \mapsto \langle v_k, x_0 \rangle$ is not identical zero on I. Thus, the set of zeros of $k \mapsto \langle v_k, x_0 \rangle$ is discrete in I. Now (i) is a consequence of Lemma 8.1.5.
(ii) follows from the fact that there exists a $k^* > 0$ such that $\sigma(A_k) \subset \overline{\mathbb{C}}_-$ for

all $k \geq k^*$, whence the set considered in (i) is bounded and therefore finite. $\square$

Remark 8.1.7

Unfortunately, previous results cannot be extended to multivariable systems. Consider, for example, the controllable and observable, minimum phase system $\dot{x}(t) = Ax(t) + Bu(t)$, $\quad y(t) = Cx(t)$, given by

$$A = \begin{bmatrix} 1 & 0 \\ 0 & -1 \end{bmatrix}, \quad B = C^T = I_2.$$

$v_k = (1, 0)^T$ is an unstable eigenvector of $A - kBC = \mathrm{diag}\{1 - k, -1 - k\}$ with eigenvalue $(1 - k)$ for all $k \in [0, 1]$. Since $\langle v_k, e_2 \rangle = 0$ for all $k \in \mathbb{R}$, Lemma 8.1.4 cannot be generalized. The same is true for Theorem 8.1.6 since the set considered in (i) respectively (ii), for $x_0 = (1, 0)^T$, is the continuum $(-\infty, 1]$ respectively $[0, 1]$.

8.2   Topological aspects

In this section, we shall consider controllable single-input, single-output, minimum phase systems of relative degree 1, that are systems of the form

$$\left. \begin{array}{rcll} \dot{x}(t) & = & Ax(t) + bu(t), & x(0) = x_0 \\ y(t) & = & cx(t), & \end{array} \right\} \tag{8.10}$$

with $(A, b, c) \in \mathbb{R}^{n \times n} \times \mathbb{R}^n \times \mathbb{R}^{1 \times n}$ and $cb \neq 0$. The result of Theorem 4.5.1 is that the adaptive feedback strategy

$$\left. \begin{array}{rcll} u(t) & = & -T_i y(t) & \text{, if } s(t) \in [T_{i-1}, T_i) \\ \dot{s}(t) & = & y(t)^2 & , s(0) = T_0 \end{array} \right\} \tag{8.11}$$

if $cb > 0$, respectively

$$\left. \begin{array}{rcll} u(t) & = & -(-1)^i T_i y(t) & \text{, ..} \ s(t) \in [T_{i-1}, T_i) \\ \dot{s}(t) & = & y(t)^2 & , s(0) = T_0 \end{array} \right\} \tag{8.12}$$

if $cb \neq 0$, applied to (8.10), under certain growth conditions of the strictly increasing sequence $T = \{T_i\}_{i \in \mathbb{N}}$ of real numbers, yields a closed-loop system with exponential decaying $x(t)$, and the switching mechanism switches only a finite number of times. It is not guaranteed that the terminal system, i.e.

$$\dot{x}(t) = \left[A - (-1)^M T_M bc \right] x(t)$$

where

$$M := \inf \left\{ i \in \mathbb{N} \mid \lim_{t \to \infty} s(t) \leq T_i \right\},$$

is exponentially stable. Now we will show that, for fixed initial condition x_0, the set of sequences of thresholds which lead to an exponentially stable terminal system is dense and that, at each time t_i where the gain T_i switches, the new trajectory

$$e^{[A-(-1)^{i+1}T_{i+1}bc](t-t_i)}x(t_i)$$

is not exponentially decaying with probability 1 whenever $[A - (-1)^{i+1}T_{i+1}bc]$ is not exponentially stable. To make this more precise, let

$$\mathcal{T} := \left\{ T = \{T_i\}_{i \in \mathbb{N}} \in \mathbb{R}^{\mathbb{N}} \mid T \quad \text{satisfies} \quad (8.13) \right\},$$

where

$$T_i > T_{i-1} \quad \text{and} \quad \lim_{i \to \infty} [T_i - T_{i-1}] = \infty \qquad (8.13)$$

and define a subspace of $\mathcal{T}$ which is relevant for the switching strategy (8.12)

$$\hat{\mathcal{T}} := \left\{ T \in \mathcal{T} \mid T \quad \text{satisfies} \quad (8.14) \right\},$$

where

$$\left. \begin{array}{rcl} \displaystyle\inf_{l \in \mathbb{N}} \frac{\sum_{i=1}^{l}(-1)^i T_i (T_i - T_{i-1})}{T_l - T_0} & = & -\infty, \\[3mm] \displaystyle\sup_{l \in \mathbb{N}} \frac{\sum_{i=1}^{l}(-1)^i T_i (T_i - T_{i-1})}{T_l - T_0} & = & +\infty. \end{array} \right\} \qquad (8.14)$$

For $\varepsilon > 0$, we define the open ball with centre $T = \{T_i\}_{i \in \mathbb{N}} \in \mathbb{R}^{\mathbb{N}}$ to be

$$B_\varepsilon(T) := \left\{ S = \{S_i\}_{i \in \mathbb{N}} \in \mathbb{R}^{\mathbb{N}} \mid |S_i - T_i| < \varepsilon \quad \text{for all} \quad i \in \mathbb{N} \right\}.$$

Using this terminology we have:
Consider a controllable, minimum phase system of the form (8.10) with fixed $x_0 \in \mathbb{R}^n$, $x_0 \neq 0$. Then the set of the sequences of thresholds $\mathcal{T}_s$ (resp. $\hat{\mathcal{T}}_s$), i.e. all sequences $T \in \mathcal{T}$ (resp. $T \in \hat{\mathcal{T}}$) so that T in combination with (8.11) (resp. (8.12)) leads to an exponentially stable terminal system, is dense in $\mathcal{T}$ (resp. $\hat{\mathcal{T}}$).

Theorem 8.2.1
Suppose (8.10) is a controllable, minimum phase system with $cb \neq 0$ and fixed $x_0 \in \mathbb{R}^n$, $x_0 \neq 0$. If $T \in \mathcal{T}$ and (8.11) (resp. $T \in \hat{\mathcal{T}}$ and (8.12)) is applied to (8.10), then for every $\varepsilon > 0$ there exists $\tilde{T} \in B_\varepsilon(T)$ such that $\tilde{T}$ leads to an exponentially stable terminal system and T and $\tilde{T}$ differ in only finitely many points.

Proof: We consider the switching strategy (8.12) only, (8.11) is simpler. Suppose the switching algorithm using the nominal sequence T leads to a terminal system $[A - (-1)^M T_M bc]$ which is not exponentially stable. Then, in particular, $s(t) \in [T_{M-1}, T_M)$ for all $t \geq t_{M-1}$, the last switch occurs at time t_{M-1}, and

$$e^{[A-(-1)^M T_M bc](t-t_{M-1})} x(t_{M-1}) \to 0 \quad \text{as} \quad t \to \infty.$$

By Theorem 8.1.6 (i), the set of $k \in \mathbb{R}$ so that

$$\sigma(A - kbc) \cap \overline{\mathbb{C}}_+ \neq \emptyset \quad \text{and} \quad \lim_{t \to \infty} e^{[A-(-1)^M T_M bc](t-t_{M-1})} x(t_{M-1}) = 0.$$

is discrete. Thus we can choose $\varepsilon' \in (0, \varepsilon)$ and S_M instead of T_M so that

$$T_{M-1} < T_M - \varepsilon' < S_M < T_M + \varepsilon' < T_{M+1},$$

and

$$e^{[A-(-1)^M S_M bc](t-t_{M-1})} x(t_{M-1}) \not\to 0 \quad \text{as} \quad t \to \infty.$$

Since (A, b, c) is detectable, see Proposition 2.1.2, $s(t)$ will leave the interval $[T_{M-1}, S_M)$. Proceeding in this way, and changing the nominal switching sequence at each switch, so that the projection of the state on the unstable subspace is nonzero, the switching strategy will, according to Theorem 4.5.1, stop after finitely many switches. Therefore, the terminal system must be exponentially stable. This completes the proof. □

An immediate consequence of Theorem 8.1.6 (i) is the following observation.

Remark 8.2.2
Suppose $T \in \mathcal{T}$ and (8.11) (resp. $T \in \hat{\mathcal{T}}$ and (8.12)) is applied to a controllable minimum phase system (8.10) with $x_0 \neq 0$ and $cb > 0$ (resp. $cb \neq 0$), then at each time

$$t_i := \inf \{t \geq t_{i-1} \mid s(t) = T_i\} \quad , t_0 := 0$$

when a new gain is chosen, the new trajectory

$$x(t; t_i) = e^{[A-T_{i+1} bc](t-t_i)} x(t_i)$$

respectively

$$\hat{x}(t; t_i) = e^{[A-(-1)^{i+1} T_{i+1} bc](t-t_i)} x(t_i)$$

satisfies

$$x(t; t_i) \not\to 0 \quad \text{resp.} \quad \hat{x}(t; t_i) \not\to 0 \quad \text{as} \quad t \to \infty$$

with probability 1 with respect to $T_{i+1} \in \mathbb{R}$ if

$$\sigma(A - T_{i+1} bc) \cap \overline{\mathbb{C}}_+ \neq \emptyset$$

respectively

$$\sigma(A - (-1)^{i+1} T_{i+1} bc) \cap \overline{\mathbb{C}}_+ \neq \emptyset.$$

If the universal adaptive stabilizer is used in series with an internal model to design a universal adaptive tracking controller, see Section 5.1, then the problem of adaptive tracking was converted into a problem of adaptive stabilizing. Thus the result of the following theorem would be expected, namely that previous topological properties are also valid for an asymptotic tracking controller.

Theorem 8.2.3
Suppose the system

$$\begin{aligned} \dot{x}(t) &= Ax(t) + bu(t), \qquad x(0) = x_0 \\ y(t) &= cx(t), \end{aligned} \right\} \tag{8.15}$$

with $(A, b, c) \in \mathbb{R}^{n \times n} \times \mathbb{R}^n \times \mathbb{R}^{1 \times n}$, $cb \neq 0$, and $x_0 \neq 0$, is controllable and minimum phase, and a class of reference signals is given by

$$\mathcal{Y}_{\text{ref}} := \left\{ y_{\text{ref}}(\cdot) \in C^\infty(\mathbb{R}, \mathbb{R}^m) \mid \alpha(\tfrac{d}{dt}) y_{\text{ref}}(t) \equiv 0 \right\},$$

where $\alpha(\cdot) \in \mathbb{R}[s]$. Choose a Hurwitz polynomial $\beta(\cdot) \in \mathbb{R}[s]$ with $\deg \beta(\cdot) = \deg \alpha(\cdot)$, and a minimal realization of $\frac{\beta(\cdot)}{\alpha(\cdot)} I_m$ as in (5.2).
Let $T \in \hat{\mathcal{T}}$. If the switching strategy

$$\begin{aligned} e(t) &= y(t) - y_{\text{ref}}(t) \\ \dot{s}(t) &= e(t)^2 && , s(0) = T_0 \\ v(t) &= -(-1)^i T_i e(t) && , \text{ if } s(t) \in [T_{i-1}, T_i) \\ \dot{\xi}(t) &= \hat{A}^* \xi(t) + \hat{B}^* v(t) && , \xi(0) = \xi_0 \in \mathbb{R}^m \\ u(t) &= \hat{C}^* \xi(t) + \hat{d} I_m v(t) \end{aligned}$$

is applied to (8.15), then the the assertions (i)-(iv) of Theorem 5.1.3 are valid and, moreover, if the sequence of thresholds T does not produce an exponentially stable terminal system $\left[\bar{A} - (-1)^M T_M \bar{B} \bar{C} \right]$, then for every $\varepsilon > 0$ there exists a $\tilde{T} \in \mathcal{B}_\varepsilon(T)$ such that $\tilde{T}$ leads to an exponentially stable terminal system and, T and $\tilde{T}$ differ in only finitely many points.

Proof: Transform the tracking problem into a stabilization problem as done in the proof of Theorem 5.1.3. Now the result follows by using analogous arguments as in the proof of Theorem 8.2.1. We omit it. □

Remark 8.2.4
Clearly, the switching strategy in Theorem 8.2.3 can be simplified by replacing $v(t)$ by $v(t) = -T_i e(t)$ if it is known in advance that $cb > 0$.
A analogous remark as in Remark 8.2.2 is valid.

8.3   Notes and References

Proposition 8.1.2 (for single-input, single-output systems), Proposition 8.1.3, and Theorem 8.2.1 have been proved in Ilchmann and Owens (1991c). Lemma 8.1.4 has been claimed partly in Ilchmann and Owens (1991c), but the proof contains several gaps. All other results are from Ilchmann (1992).

Townley (1992) investigated the same universal adaptive stabilization strategy and proved that for fixed $k_0 \in \mathbb{R}$, the set of $x_0 \in \mathbb{R}^n$ so that the terminal system is exponentially stable is open and dense with complement which is nowhere dense.

References

ÅSTRÖM, K.J. (1987) Adaptive feedback control, *Proceedings of the IEEE* 75, 185-217

ANDERSON, B.D.O. and S. VONGPANITLERD (1973) *Network Analysis and Synthesis - A Modern Systems Theory Approach*, Prentice-Hall, Englewood Cliffs

BAR-KANA, I. and H. KAUFMAN (1985) Global stability and performance of a simplified adaptive algorithm, *Int. J. Control* 42, 1491-1505

BYRNES, C.I. (1987) Adaptive stabilization of infinite dimensinal linear systems, *Proc. 26th Conf. on Decision and Control*, Los Angeles, 1435-1440

BYRNES, C.I., U. HELMKE and A.S. MORSE (1986) Necessary conditions in adaptive control, pp. 3-14 in *Modelling Identification and Robust Control* (C.I. Byrnes and A. Linquist, eds.), Elsevier Science Pubs., North-Holland

BYRNES, C.I. and A. ISIDORI (1986) Asymptotic expansions, root-loci and the global stability of nonlinear feedback systems, pp. 159-179 in *Algebraic and Geometric Methods in Nonlinear Control Theory* (M. Fliess and M. Hazewinkel, eds.), D.Reidel Publishing Company

BYRNES, C.I. and J.C. WILLEMS (1984) Adaptive stabilization of multivariable linear systems, *Proc. 23rd Conf. on Decision and Control*, Las Vegas, 1574-1577

COPPEL, W.A. (1974) Matrices of rational functions, *Bull. Austral. Math. Soc.* 11, 89-113

CORLESS, M. (1988) First order adaptive controllers for systems which are stabilizable via high gain output feedback, pp. 13-16 in *Analysis and Control of Nonlinear Systems* (C.I. Byrnes, C.F. Martin and R.E. Seaks, eds.), Elsevier Science Pubs., North-Holland

CORLESS, M. (1991) Simple adaptive controllers for systems which are stabilizable via high gain feedback, *IMA Journal of Math. Control and Information* 8, 379-387

CORLESS, M. and E.P. RYAN (1992) Adaptive control of a class of nonlinearly perturbed linear systems of relative degree two, Preprint

DAHLEH, M. (1988) Sufficient information for the adaptive stabilization of delay systems, *Syst. Control Lett.* 11, 357-363

DAHLEH, M. (1989) Generalizations of Tychonov's theorem with applications to adaptive control of SISO delay systems, *Syst. Control Lett.* 13, 421-427

DAHLEH, M. and W.E. HOPKINS, JR. (1986) Adaptive stabilization of single-input single-output delay systems, *IEEE Trans. Aut. Control* 31, 577-579

GANTMACHER, F.R. (1959), *The Theory of Matrices Vol.1 & 2*, Chelsea Publ., New York

GOHBERG, I., P. LANCASTER and L. RODMAN (1982), *Matrix Polynomials*, New York, Academic Press

HALE, J. (1977) *Theory of Functional Differential Equations*, Springer-Verlag, New York

HELMKE, U. and D. PRÄTZEL-WOLTERS (1988) Robustness properties of universal adaptive stabilizers for first order systems, *Int. J. Control* 48, 1153-1182

HELMKE, U., D. PRÄTZEL-WOLTERS and S. SCHMID (1990) Adaptive tracking for scalar minimum phase systems, pp. 101-118 in *Control of Uncertain Systems* (D. Hinrichsen and B. Mårtensson, eds.), Boston, Birkhäuser

HEYMANN, M., J.H. LEWIS and G. MEYER (1985) Remarks on the adaptive control of linear plants with unknown high-frequency gain, *Syst. Control Lett.* 5, 357-362

HICKS, A.C. and S. TOWNLEY (1992) On exact solutions of differential equations arising in universal adaptive stabilization, to appear in *Syst. Control Lett.*

ILCHMANN A. (1991) Non-identifier-based adaptive control of dynamical systems: a survey. *IMA Journal of Math. Control and Inf.* 8, 321-366

ILCHMANN A. (1992) Adaptive controllers and root-loci of minimum phase systems, submitted to *Dynamics and Control*

ILCHMANN A. and H. LOGEMANN (1992) High-gain adaptive stabilization of multivariable linear systems - revisited, *Syst. Control Lett.* 18, 355-364

ILCHMANN A. and D.H. OWENS (1990) Adaptive stabilization with exponential decay, *Syst. Control Lett.* 14, 437-443

ILCHMANN, A. and D.H. OWENS (1991) Exponential stabilization using non-differential gain adaptation, *IMA Journal of Math. Control and Information* 7, 339-349

ILCHMANN A. and D.H. OWENS (1991a) Threshold switching functions in high-gain adaptive control, *IMA Journal of Math. Control and Information* 8, 409-429

ILCHMANN A. and D.H. OWENS (1991b) Robust universal adaptive stabilization in the presence of nonlinearities, *Proc. of the First European Control Conf.*, Grenoble, 2580-2585

ILCHMANN A. and D.H. OWENS (1991c) Exponential stabilization using piecewise constant gain adaptation, *Proc. 30-th IEEE Conf. on Decision and Control*, Brighton, 83-84

ILCHMANN A. and D.H. OWENS (1992) Adaptive exponential tracking for nonlinearly perturbed minimum phase systems, submitted to *Control-Theory and Advanced Technology*

ILCHMANN, A., D.H. OWENS and D. PRÄTZEL-WOLTERS (1987) High gain robust adaptive controllers for multivariable systems, *Syst. Control Lett.* 8, 397-404

ILCHMANN A. und E.P. RYAN (1992) Adaptive tracking for nonlinear systems in the presence of noise, submitted to *Automatica*

ILCHMANN A. und S. TOWNLEY (1992) Simple adaptive stabilization of high-gain stabilizable systems, to appear in *Syst. Control Lett.*

KAILATH, T. (1980) *Linear Systems* , Prentice-Hall, Englewood Cliffs

KATO, T. (1976) *Perturbation Theory for Linear Operators*, Berlin, Springer-Verlag

KHALIL, H.K. and A. SABERI (1987) Adaptive stabilization of a class of nonlinear systems using high-gain feedback, *IEEE Trans. Aut. Control* 32, 1031-1035

KOBAYASHI, T. (1987) Global adaptive stabilization of infinite-dimensional systems, *Syst. Control Lett.* 9, 215-223

KOKOTOVIĆ, P.V. (1984) Applications of singular perturbation techniques to control problems, *SIAM Review* 26, 501-551

LASALLE, J.P. (1976) Stability of nonautonomous systems, *Nonlinear Analysis, Theory, Methods & Appl.* 1, 83-91

LOGEMANN, H. (1990) Adaptive exponential stabilization for a class of nonlinear retarded processes, *Mathematics of Control, Signals, and Systems* 3, 255-269

LOGEMANN H. und A. ILCHMANN (1991) An adaptive servomechanism for a class of infinite-dimensional systems, Report No.244, Institut für Dynamische Systeme, Universität Bremen, submitted to *SIAM J. on Contr.*

LOGEMANN, H. and B. MÅRTENSSON (1991) Adaptive stabilization of infinite dimensional systems, to appear in *IEEE Trans. Aut. Control*

LOGEMANN, H. and D.H. OWENS (1988) Input-output theory of high-gain adaptive stabilization of infinite dimensional systems with nonlinearities, *Int. J. Adap. Control and Signal Processing* 2, 193-216

LOGEMANN, H. and D.H. OWENS (1988) Robust and adaptive high-gain control of infinite-dimensional systems, pp. 35-44 in *Analysis and Control of Nonlinear Systems* (C.I. Byrnes, C.F. Martin and R.E. Seaks, eds.), Elsevier Science Pubs., North-Holland

LOGEMANN, H. and H. ZWART (1991) Some remarks on adaptive stabilization of infinite-dimensional systems, *Syst. Control Lett.* 16, 199-207

MACFARLANE, A.G.J. and N. KARCANIAS (1978) Relationships between state-space and frequency-response concepts, *Proc. 7-th IFAC World Congress*, Helsinki

MACFARLANE, A.G.J. and I. POSTLETHWAITE (1977) The generalized Nyquist stability criterion and multivariable root-loci, *Int. J. Control* 25, 81-127

MACIEJOWSKI, J.M. (1989) *Multivariable Feedback Design*, Addison-Wesley, Wokingham, England

MAREELS, I. (1984) A simple selftuning controller for stably invertible systems, *Syst. Control Lett.* 4, 5-16

MÅRTENSSON, B. (1985) The order of any stabilizing regulator is sufficient a priori information for adaptive stabilization, *Syst. Control Lett.* 6, 87-91

MÅRTENSSON, B. (1986) *Adaptive stabilization* ; Thesis, Lund Institute of Technology, Lund, Sweden

MÅRTENSSON, B. (1987) Adaptive stabilization of multivariable linear systems, *Contemporary Mathematics* 68, 191-225

MÅRTENSSON, B. (1990) Remarks on adaptive stabilization of first order nonlinear systems, *Syst. Control Lett.* 14, 1-7

MÅRTENSSON, B. (1991) The unmixing problem, *IMA Journal of Math. Control and Information* 8, 367-377

MILLER, D.E. and E.J. DAVISON (1989), An adaptive controller which provides Lyapunov stability, *IEEE Trans. Aut. Control* 34, 599-609

MILLER, D.E. and E.J. DAVISON (1991) An adaptive controller which provides an arbitrarily good transient and steady-state response, *IEEE Trans. Aut. Control* 36, 68-81

MILLER, D.E. and E.J. DAVISON (1991a), An adaptive tracking problem, Systems Control Group Report 9113, Dept. of Electr. Engg., University of Toronto

MORSE, A.S. (1983) Recent problems in parameter adaptive control, pp. 733-740 in *Outils et Modèles Mathématiques pour l'Automatique, l'Analyse de Systèmes et le Traitment du Signal* (I.D. Landau, Ed.), (Editions du CNRS 3, Paris)

MORSE, A.S. (1984) An adaptive control for globally stabilizing linear systems with unknown high-frequency gains, pp. 58-68 in *Lect. Notes Control Inf. Sci.* 62, Springer-Verlag, Berlin

MORSE, A.S. (1986) Simple algorithms for adaptive stabilization, in: *Proc. ISSA, Conf. on Modelling and Adaptive Control*, Sopron, Hungary, and pp. 254-264 in *Lect. Notes Control Inf. Sci.* 105, Springer-Verlag, Berlin,(1988)

MORSE, A.S. (1987a) *High-gain feedback algorithms for adaptive stabilization* ; Proc. of the 5th Yale Workshop on Appls. of Adaptive Systems Theory, 13-18

MUDGETT, D.R. and A.S. MORSE (1989) A smooth, high-gain adaptive stabilizer for linear systems with known relative degree, *Proc. American Control Conference*, Pittsburgh, PA, 21-26

NARENDRA, K.S. (1991) The maturing of adaptive control, pp. 3-36 in *Lect. Notes in Control and Inf. Sciences* 160, Springer-Verlag, Berlin

NARENDRA, K.S. and J.H. TAYLOR (1973) *Frequency Domain Criteria for Absolute Stability*, Academic Press, New York

NIKITIN, S. and SCHMID, S. (1990) *Universal adaptive stabilizers for one-dimensional nonlinear systems* ; Report 44, Fachbereich Mathematik, Universität Kaiserslautern

NUSSBAUM, R.D. (1983) Some remarks on a conjecture in parameter adaptive control, *Syst. Control Lett.*, 3, 243-246

OGATA, K. (1990) *Modern Control Engineering*, second edition, Prentice-Hall, Englewood Cliffs, N.Y.

OWENS D.H., A. CHOTAI und A. ABIRI (1984) Parametrization and approximation methods in feedback theory with applications in high-gain, fast-sampling, and cheap-optimal control, *IMA Journal of Math. Control and Inf.* 1, 147-171

OWENS D.H., D. PRÄTZEL-WOLTERS und A. ILCHMANN (1987) Positive real structure and high gain adaptive stabilization, *IMA Journal of Math. Control and Inf.* 4, 167-181

PRÄTZEL-WOLTERS D., D.H. OWENS und A. ILCHMANN (1989) Robust adaptive stabilization by high gain and switching, *Int. J. Control* 49, 1861-1868

PRÄTZEL-WOLTERS, D. und R. REINKE (1991) Discrete positive real systems and high gain stability, Report 63, Dept. of Mathematics, University of Kaiserslautern

PUGH, A.C. and P.A. RATCLIFFE (1981) Infinite pole and zero considerations for system transfer function matrices, pp. 51-64 in *3rd IMA Conf. on Control Theory*, Sheffield

RYAN, E.P. (1988) Adaptive stabilization of a class of uncertain nonlinear systems: A differential inclusion approach, *Syst. Control Lett.* 10, 95-101

RYAN, E.P. (1990) Discontinuous feedback and universal adaptive stabilization, pp. 245-258 in *Control of Uncertain Systems* (D. Hinrichsen and B. Mårtensson, eds.), Boston, Birkhäuser

RYAN, E.P. (1991) A universal adaptive stabilizer for a class of nonlinear systems, *Syst. Control Lett.* 16, 209-218

RYAN, E.P. (1991a) Adaptive stabilization of multi-input nonlinear systems, to appear in *Int. J. of Robust and Nonlinear Control*

RYAN, E.P. (1992) Universal $W^{1,\infty}$-tracking for a class of nonlinear systems, *Syst. Control Lett.* 18, 201-210

RYAN, E.P. (1992a) A nonlinear universal servomechanism, preprint

SABERI, A. and Z. LIN (1990) Adaptive high-gain stabilization of 'minimum-phase' nonlinear systems, *Proc. 29th IEEE Conf. on Decision of Control*, Honolulu, Hawaii, 3417-3422

SCHMID, S. (1991) *Adaptive Synchronization of Interconnected Systems*, Ph.D. Thesis, Dept. of Mathematics, University of Kaiserslautern, Germany

TAO, G. and P. IOANNOU (1988) Stricly positive real matrices and the Lefschetz-Kalman-Yakubovich Lemma, *IEEE Trans. Aut. Control* 33, 1183-1185

TOWNLEY, S. (1992) Topological aspects of universal adaptive stabilization, Dept. of Maths., University of Exeter, England

TOWNLEY, S. and D.H. OWENS (1991) A note an the problem of multivariable adaptive tracking, *IMA Journal of Math. Control and Information* 8, 389-395

VIDYASAGAR, M. (1978) *Nonlinear Systems Analysis*, Prentice-Hall, Englewood Cliffs

WILLEMS, J.C. and C. I. BYRNES (1984) Global adaptive stabilization in the absence of information on the sign of the high frequency gain, pp. 49-57 in *Lect. Notes in Control and Inf. Sciences* 62, Springer-Verlag, Berlin

WONHAM, W.M. (1979) *Linear Multivariable Control: a Geometric Approach*, Springer-Verlag, New York

WEN, J.T. (1988) Time domain and frequency domain conditions for strict positive realness, *IEEE Trans. Aut. Contr.* 33, 988-992

ZHU, X.-J. (1989) A finite spectrum unmixing set for $GL_3(\mathbb{R})$, pp. 403-410 in *Computation and Control* (K. Bowers and J. Lund, eds.), Boston, Birkhäuser

Index

Lecture Notes in Control and Information Sciences

Edited by M. Thoma and A. Wyner